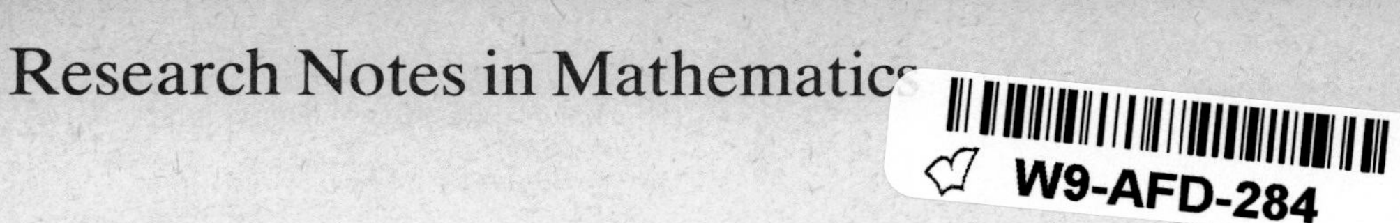

Research Notes in Mathematics

Submission of proposals for consideration
Suggestions for publication, in the form of outlines and representative samples, are invited by the editorial board for assessment. Intending authors should contact either the main editor or another member of the editorial board, citing the relevant AMS subject classifications. Refereeing is by members of the board and other mathematical authorities in the topic concerned, located throughout the world.

Preparation of accepted manuscripts
On acceptance of a proposal, the publisher will supply full instructions for the preparation of manuscripts in a form suitable for direct photo-lithographic reproduction. Specially printed grid sheets are provided and a contribution is offered by the publisher towards the cost of typing.

Illustrations should be prepared by the authors, ready for direct reproduction without further improvement. The use of hand-drawn symbols should be avoided wherever possible, in order to maintain maximum clarity of the text.

The publisher will be pleased to give any guidance necessary during the preparation of a typescript, and will be happy to answer any queries.

Important note
In order to avoid later retyping, intending authors are strongly urged not to begin final preparation of a typescript before receiving the publisher's guidelines and special paper. In this way it is hoped to preserve the uniform appearance of the series.

Titles in this series

1 Improperly posed boundary value problems
A Carasso and A P Stone
2 Lie algebras generated by finite dimensional ideals
I N Stewart
3 Bifurcation problems in nonlinear elasticity
R W Dickey
4 Partial differential equations in the complex domain
D L Colton
5 Quasilinear hyperbolic systems and waves
A Jeffrey
6 Solution of boundary value problems by the method of integral operators
D L Colton
7 Taylor expansions and catastrophes
T Poston and I N Stewart
8 Function theoretic methods in differential equations
R P Gilbert and R J Weinacht
9 Differential topology with a view to applications
D R J Chillingworth
10 Characteristic classes of foliations
H V Pittie
11 Stochastic integration and generalized martingales
A U Kussmaul
12 Zeta-functions: An introduction to algebraic geometry
A D Thomas
13 Explicit *a priori* inequalities with applications to boundary value problems
V G Sigillito
14 Nonlinear diffusion
W E Fitzgibbon III and H F Walker
15 Unsolved problems concerning lattice points
J Hammer
16 Edge-colourings of graphs
S Fiorini and R J Wilson
17 Nonlinear analysis and mechanics: Heriot-Watt Symposium Volume I
R J Knops
18 Actions of fine abelian groups
C Kosniowski
19 Closed graph theorems and webbed spaces
M De Wilde
20 Singular perturbation techniques applied to integro-differential equations
H Grabmüller
21 Retarded functional differential equations: A global point of view
S E A Mohammed
22 Multiparameter spectral theory in Hilbert space
B D Sleeman
24 Mathematical modelling techniques
R Aris
25 Singular points of smooth mappings
C G Gibson
26 Nonlinear evolution equations solvable by the spectral transform
F Calogero
27 Nonlinear analysis and mechanics: Heriot-Watt Symposium Volume II
R J Knops
28 Constructive functional analysis
D S Bridges
29 Elongational flows: Aspects of the behaviour of model elasticoviscous fluids
C J S Petrie
30 Nonlinear analysis and mechanics: Heriot-Watt Symposium Volume III
R J Knops
31 Fractional calculus and integral transforms of generalized functions
A C McBride
32 Complex manifold techniques in theoretical physics
D E Lerner and P D Sommers
33 Hilbert's third problem: scissors congruence
C-H Sah
34 Graph theory and combinatorics
R J Wilson
35 The Tricomi equation with applications to the theory of plane transonic flow
A R Manwell
36 Abstract differential equations
S D Zaidman
37 Advances in twistor theory
L P Hughston and R S Ward
38 Operator theory and functional analysis
I Erdelyi
39 Nonlinear analysis and mechanics: Heriot-Watt Symposium Volume IV
R J Knops
40 Singular systems of differential equations
S L Campbell
41 N-dimensional crystallography
R L E Schwarzenberger
42 Nonlinear partial differential equations in physical problems
D Graffi
43 Shifts and periodicity for right invertible operators
D Przeworska-Rolewicz
44 Rings with chain conditions
A W Chatters and C R Hajarnavis
45 Moduli, deformations and classifications of compact complex manifolds
D Sundararaman
46 Nonlinear problems of analysis in geometry and mechanics
M Atteia, D Bancel and I Gumowski
47 Algorithmic methods in optimal control
W A Gruver and E Sachs
48 Abstract Cauchy problems and functional differential equations
F Kappel and W Schappacher
49 Sequence spaces
W H Ruckle
50 Recent contributions to nonlinear partial differential equations
H Berestycki and H Brezis
51 Subnormal operators
J B Conway
52 Wave propagation in viscoelastic media
F Mainardi
53 Nonlinear partial differential equations and their applications: Collège de France Seminar. Volume I
H Brezis and J L Lions
54 Geometry of Coxeter groups
H Hiller
55 Cusps of Gauss mappings
T Banchoff, T Gaffney and C McCrory

56 An approach to algebraic K-theory
A J Berrick
57 Convex analysis and optimization
J-P Aubin and R B Vintner
58 Convex analysis with applications in the differentiation of convex functions
J R Giles
59 Weak and variational methods for moving boundary problems
C M Elliott and J R Ockendon
60 Nonlinear partial differential equations and their applications: Collège de France Seminar. Volume II
H Brezis and J L Lions
61 Singular systems of differential equations II
S L Campbell
62 Rates of convergence in the central limit theorem
Peter Hall
63 Solution of differential equations by means of one-parameter groups
J M Hill
64 Hankel operators on Hilbert space
S C Power
65 Schrödinger-type operators with continuous spectra
M S P Eastham and H Kalf
66 Recent applications of generalized inverses
S L Campbell
67 Riesz and Fredholm theory in Banach algebra
B A Barnes, G J Murphy, M R F Smyth and T T West
68 Evolution equations and their applications
F Kappel and W Schappacher
69 Generalized solutions of Hamilton-Jacobi equations
P L Lions
70 Nonlinear partial differential equations and their applications: Collège de France Seminar. Volume III
H Brezis and J L Lions
71 Spectral theory and wave operators for the Schrödinger equation
A M Berthier
72 Approximation of Hilbert space operators I
D A Herrero
73 Vector valued Nevanlinna Theory
H J W Ziegler
74 Instability, nonexistence and weighted energy methods in fluid dynamics and related theories
B Straughan
75 Local bifurcation and symmetry
A Vanderbauwhede
76 Clifford analysis
F Brackx, R Delanghe and F Sommen
77 Nonlinear equivalence, reduction of PDEs to ODEs and fast convergent numerical methods
E E Rosinger
78 Free boundary problems, theory and applications. Volume I
A Fasano and M Primicerio
79 Free boundary problems, theory and applications. Volume II
A Fasano and M Primicerio
80 Symplectic geometry
A Crumeyrolle and J Grifone
81 An algorithmic analysis of a communication model with retransmission of flawed messages
D M Lucantoni
82 Geometric games and their applications
W H Ruckle
83 Additive groups of rings
S Feigelstock
84 Nonlinear partial differential equations and their applications: Collège de France Seminar. Volume IV
H Brezis and J L Lions
85 Multiplicative functionals on topological algebras
T Husain
86 Hamilton-Jacobi equations in Hilbert spaces
V Barbu and G Da Prato
87 Harmonic maps with symmetry, harmonic morphisms and deformations of metrics
P Baird
88 Similarity solutions of nonlinear partial differential equations
L Dresner
89 Contributions to nonlinear partial differential equations
C Bardos, A Damlamian, J I Díaz and J Hernández
90 Banach and Hilbert spaces of vector-valued functions
J Burbea and P Masani
91 Control and observation of neutral systems
D Salamon
92 Banach bundles, Banach modules and automorphisms of C*-algebras
M J Dupré and R M Gillette
93 Nonlinear partial differential equations and their applications: Collège de France Seminar. Volume V
H Brezis and J L Lions

Banach bundles, Banach modules and automorphisms of C*-algebras

M J Dupré & R M Gillette
Tulane University/Montana State University

Banach bundles, Banach modules and automorphisms of C*-algebras

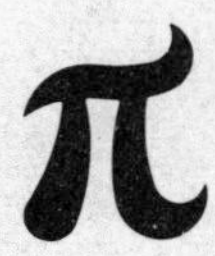

Pitman Advanced Publishing Program
BOSTON · LONDON · MELBOURNE

PITMAN BOOKS LIMITED
128 Long Acre, London WC2E 9AN

PITMAN PUBLISHING INC
1020 Plain Street, Marshfield, Massachusetts 02050

Associated Companies
Pitman Publishing Pty Ltd, Melbourne
Pitman Publishing New Zealand Ltd, Wellington
Copp Clark Pitman, Toronto

First published 1983

AMS Subject Classifications: (main) 46L05
(subsidiary) 55R05, 55R10, 55R15

Library of Congress Cataloging in Publication Data

Dupré, Maurice J.
Banach bundles, Banach modules and automorphisms of C*-algebras.

(Research notes in mathematics; 92)
Includes bibliographical references and index.
1. C*-algebras. 2. Fiber spaces (Mathematics)
3. Fiber bundles (Mathematics) I. Gillette, R. M.
II. Title. III. Series.
QA326.D86 1983 512′.55 83-11656
ISBN 0-273-08626-X

British Library Cataloguing in Publication Data

Dupré, M. J.
Banach bundles, Banach modules and automorphisms of C*-algebras.—(Research notes in mathematics; 92)
1. C*-algebras
I. Title II. Gillette, R. M. III. Series
512′.55 QA326

ISBN 0-273-08626-X

Reproduced and printed by photolithography
in Great Britain by Biddles Ltd, Guildford

Contents

The research of M J Dupré at the Department of Mathematics, Tulane University, was partially supported by the National Science Foundation.

Preface

In recent years the theory of bundles or fibre spaces has found an increasing number of applications in functional analysis, C*-algebra theory, and Banach space theory. However, unlike the bundles topologists usually like to deal with, the bundles that arise in analysis are often not locally trivial. Even though the Banach bundles studied here satisfy the homotopy lifting property, this is misleading since their associated sphere bundles do not. Thus the theory here originally developed with a flavor quite different from that employed by topologists, and because of the added complications posed by the failure of local triviality, even the basic definitions have only recently come to be agreed upon by mathematicians in the field. The upshot is that there are basically two types of Banach bundles to be considered and although their theories run parallel, there are important differences. The present book grew out of a research article originally written in 1979 in which the authors needed to apply both theories of Banach bundles somewhat simultaneously. Thus our first two chapters form a fairly concise introduction to the basics of bundle theory as needed by the functional analyst and our last two chapters give an application of the theory to the problem of classifying certain automorphisms of C*-algebras.

Introduction

In this monograph we introduce a Picard group $Pic(A,Z)$ for an arbitrary unital C^*-algebra A with center Z, and we establish an exact sequence of groups

$$1 \to \text{Inn } A \to \text{LInn } A \to Pic(A,Z) \to 1 ,$$

in which LInn A is a certain subgroup of the group of automorphisms of A which commute with the action of Z (see Theorem 3.12). This exact sequence generalizes results of Phillips and Raeburn [36] which in turn generalize the theorems of Lance [33] and Smith [40] characterizing the group of outer $C(X)$-automorphisms of $A = C(X,B(H))$ as the second integral Čech cohomology group $\check{H}^2(X;\mathbb{Z})$ when X is compact and metrizable. The principal devices used here are the representation of A as the section algebra of a C^*-bundle ξ_A over a compact space X, and an accompanying representation theory for the Hilbert $C(X)$-submodules of A. Each $C(X)$-automorphism α of A determines, up to isomorphism, a Hilbert $C(X)$-module

$$M_\alpha = \{m \in A : \alpha(a)m = ma \quad \text{for all} \quad a \in A\} ,$$

which then determines, again up to isomorphism, a Hilbert bundle ξ_α over X having fibre dimension everywhere ≤ 1. Taking $X = \hat{Z}$ (the spectrum of $Z = Z(A)$), the Picard group is isomorphic to the group of isomorphism classes of those Hilbert bundles of the form ξ_α which have fibre dimension everywhere equal to 1. Thus $Pic(A,Z)$ is essentially a subgroup of the group of complex line bundles over the spectrum of the center of A. In terms of automorphisms of ξ_A, LInn A turns out to be the group of locally inner automorphisms of A (see Proposition 3.11).

Chapter 1 is an introduction to the theory of Banach bundles, designed for convenient access to both J.M.G. Fell's approach which utilizes continuous section norms, and the approach due to K. H. Hofmann which utilizes upper semicontinuous norms. The development here is somewhat novel, being centered around Proposition 1.2, an easy result which allows rapid development of the basic properties of Banach bundles and Banach bundle maps, and which is used repeatedly in Chapters 2 and 3. As many of the results for general Banach bundles remain somewhat obscurely buried in the literature, we have taken this development of the basics far enough to make our treatment essentially self-contained.

In Chapter 2 the representation theory of C^*-algebras and their associated modules is reviewed in a presentation which incorporates recent simplifying developments in sectional representation theory (due mainly to K. H. Hofmann and his student J. Varela) and also takes advantage of the developments in Chapter 1. In particular, we give a quick proof of the main result of J. Dauns and K. H. Hofmann for C^*-algebras [3] which asserts that any C^*-algebra is *-isomorphic to the C^*-algebra of sections of a C^*-bundle over the spectrum of its centroid (see Theorem 2.4). The most rapid route to this result is obtained by skipping directly from Proposition 1.3 to Chapter 2. As we must deal with Banach bundle maps and Banach modules, we have given a careful development here and in our Theorem 2.6 we obtain a slight extension of Theorem 2.4 which can be technically useful, the result here being more general even for C^*-algebras. We also give a result (Proposition 2.7) which illuminates the structure of the spectrum of an arbitrary C^*-algebra of sections, and

which gives necessary and sufficient conditions that the representing C^*-bundle have continuous norm. Further on in Chapter 2 we prove a Stone-Weierstrass type theorem useful for C*-algebras of sections, as an application of the Stone-Weierstrass-Glimm theorem (see Proposition 2.18). Part of this result had been known before [18, Theorem 1.4].
We conclude Chapter 2 with a proof of Swan's theorem using our Theorem 2.6.

Chapter 3 contains the key definitions and results of this monograph. Here we associate Hilbert Z-submodules of A to Z-automorphisms of A and then apply the representation theory of Chapter 2 to establish our main result (Theorem 3.12). A crucial step along the way (Corollary 3.4) establishes that the bundles associated with Hilbert Z-submodules of A have fibre dimension everywhere ≤ 1. The collection of all isomorphism classes of such bundles over the spectrum of Z constitutes a partially ordered semigroup $H(A)$ which figures in an exact sequence (see Theorem 3.6):

$$1 \longrightarrow \text{Inn } A \longrightarrow \text{Aut}_Z A \longrightarrow H(A) .$$

The group Pic(A,Z) is isomorphic to the subgroup of $H(A)$ consisting of the line bundles in $H(A)$.

In Chapter 4 we study the group Pic(A,Z). We are able to characterize the elements of $\text{Pic } Z = \check{H}^2(\hat{Z};\mathbb{Z})$ which belong to Pic(A,Z). We then use this to show, in certain situations more general than those discussed in [36], that in fact Pic(A,Z) = Pic Z.

As our first two sections are mainly of an expository nature, some historical remarks are in order. The main result of [3] for C*-algebras is the non-commutative generalization of the Gelfand-Naimark theorem which states that if A is any C*-algebra, then $A \cong \Gamma(\xi_A)$ where ξ_A is a C*-bundle over the spectrum of the centroid of A and $\Gamma(\xi_A)$ denotes the set of all continuous sections of A. We can call this result the Dauns-Hofmann-Gelfand-Naimark (DHGN) theorem, and we obtain this result in our Theorem 2.4. In the original development of [3], the theorem was proved only after a lengthy development of a more general theory applicable to a larger class of Banach algebras and abstract rings so as to include sheaf representations of rings as another special case. The resulting general objects known as _uniform fields_ are more general than necessary for the DHGN theorem alone, and part of our economy of treatment results from the hindsight of knowing what type of bundle is going to be required for the theorem and restricting attention to these objects which we call Banach bundles, following Fell [19] and Hofmann [26]. Also, in the original treatment of [3], the first candidate for the base space of the bundle ξ_A was the space of primitive ideals of A, denoted Prim A, given the hull-kernel topology. Then by successive modifications, Hausdorffization, complete regularization, and compactification necessary to build ξ_A, the space Prim A gradually gets modified and becomes the spectrum of the centroid of A. Along this route, a crucial role was played by the result proved fairly early in [3] that A is a C_b(Prim A)-module so that $(fa) + P = f(P)a + P$ for $f \in C_b(\text{Prim } A)$ and $a \in A$. It is a curious fact that this result (which is not even mentioned in the introduction of [3]), rather than the DHGN, has become known as the Dauns-Hofmann theorem and consequently some readers might mistake this module result for our aim here. In fact our development and proof of the DHGN theorem makes no use of the Dauns-Hofmann theorem at all and so we will not prove it here. We will use it for some of our remarks in Chapter 2 following the DHGN theorem, but only in isolated examples of an illustrative nature. The proof of the Dauns-Hofmann theorem given in [17] is so simple that we see no reason not to refer the interested reader there.

Our later developments in Chapters 3 and 4 depend heavily on the theory of Hilbert bundles and the equivalence between Hilbert modules and Hilbert bundles as given in Theorem 2.6. Recently, the theory of Hilbert modules over general non-commutative C*-algebras has played an important role in general Ext and K-theory for C*-algebras, as well as in the theory of induced representations for C*-algebras [31, 37, 39]. The general theory of Hilbert modules is not really new and goes back at least to [28] and [35] over non-commutative C*-algebras, and in the commutative case as far back as [22] and [30]. Of course the correspondence between Hilbert modules and (F)Hilbert bundles which results as part of Theorem 2.6 has been well known for quite some time and appears for instance in [6]. Because of this correspondence, it is natural to consider a Hilbert module over a non-commutative C*-algebra as a sort of non-commutative Hilbert bundle. With this in mind, many of the results on the structure of Hilbert bundles have natural non-commutative generalizations. In particular, a key result of [31] is the Kasparov stabilization theorem for separable Hilbert modules, generalizing the result for the commutative case first obtained in [6] which in turn played an important role in [37]. In fact the analogy goes much further as shown in [15] where essentially all the main trivialization theorems for Hilbert bundles of [6] are generalized to the non-commutative case. Moreover, in [14] the theory of pullbacks is developed and a non-abelian section extension theorem is obtained for Hilbert modules. The pullback theory for Hilbert modules shows clearly the intimate relationship between induced representation theory [39] and ordinary pullbacks of bundles.

Finally we would like to express our gratitude to the referee whose efforts led to several improvements in our manuscript.

1 Banach bundles

In this section we shall present a rapid review of Banach bundle theories which will serve a twofold purpose. First, there are two different definitions of Banach bundle currently in use, one of which [19] is more restrictive and thus more manageable, while the other [26,27] is more general and thus more accommodating for purposes of sectional representation. Although both are outgrowths of the theory of continuous fields invented by R. Godement [22] and applied in [23] and [30], the definition given by J.M.G. Fell [19] is actually a fibre space characterization of the original continuous fields of Banach spaces considered in [22], whereas the definition in [26,27] is the analogous definition needed to characterize certain of the more general continuous fields considered in [3]. As we must consider both types of Banach bundles and corresponding C^*-bundles in what follows, we shall need to be clear on the factual similarities and differences. In particular, in Chapter 2, we shall give a characterization of C^*-algebras for which the natural representation considered in [3] results in a C^*-bundle of the more restrictive type [13]. Second, recent developments of the sectional representation theory allow a very simple development of the representation of C^*-algebras and modules as spaces of sections. This development seems still to be obscurely buried in the literature and thus accessible only to the reader already fairly familiar with the Banach bundle territory. Our review will thus serve to simplify the theory of Banach bundles of the more general type as employed in [3]. In order to proceed further we must set down some notation and make some definitions.

By a <u>bundle</u> or <u>fibre space</u> we simply mean a triple $\xi = (p,E,X)$ in which $p: E \to X$ is a continuous map. As usual, p is termed the <u>projection</u> of ξ, X the <u>base</u> of ξ, E the <u>total</u> or <u>bundle</u> space of ξ, and for $x \in X$, we call $\xi(x) = E_x = p^{-1}(x)$ the <u>fibre</u> of ξ over x. By a <u>fibre bundle</u> we mean the usual notion as defined by N. Steenrod in [41]. The underlying set of a space Y is denoted $|Y|$. In this chapter, we can take either $\mathbb{R}$ or $\mathbb{C}$ as the field of scalars which we denote by $\mathbb{K}$. We write $\phi : M \cong N$, when M and N are Banach spaces, to mean ϕ is an isometric isomorphism (onto N).

DEFINITION 1.1. Let τ be a topology on $|\mathbb{R}|$. By a (τ) <u>Banach bundle</u> we mean a bundle $\xi = (p,E,X)$ such that

(1) p is open and surjective;

(2) for each $x \in X$, $|\xi(x)|$ is the underlying set of a Banach space;

(3) the maps $\mathbb{K} \times E \xrightarrow{\cdot} E$, $E \times_X E \xrightarrow{+} E$, $E \xrightarrow{\|\cdot\|} (|\mathbb{R}|),\tau)$, given in each fibre by scalar multiplication, addition, and the norm, respectively, are continuous;

(4) if $0(x) \in W$, where $0(x)$ is the zero of $\xi(x)$ and W is open in E, then there is an $\varepsilon > 0$ and an open neighborhood U of x in X such that

$$\{b \in p^{-1}(U) : \|b\| < \varepsilon\} \subset W.$$

We will be interested in only two topologies on $|\mathbb{R}|$, namely the usual topology, τ_u, and the topology τ^+ whose basis consists of all intervals of the form $(-\infty,\varepsilon)$.

By an (F) *Banach bundle* we mean a (τ_u) Banach bundle, whereas by an (H) *Banach bundle* we mean a (τ^+) Banach bundle. The (F) Banach bundles were first defined by J.M.G. Fell in [19] as the fibre space theoretic analog of the notion of a continuous field of Banach spaces defined by R. Godement in [22]. The (H) Banach bundles were first defined by K. H. Hofmann as the corresponding analog of the slightly more general notion of uniform field [3,26]. Obviously the only difference is the norm continuity requirement - the ordinary continuity of an (F) Banach bundle is relaxed to upper semicontinuity for an (H) Banach bundle. Any assertion about Banach bundles is meant to be valid for either type of Banach bundle. Generally speaking, the notation and terminology will be consistent with that in [13].

By a *fibre set* we mean a triple $\xi = (p,E,X)$ in which X is a space, E is a set, and $p : E \to X$ is a function. The function p is termed the *projection* of ξ, E is the *total set* of ξ, X is the *base space* of ξ, and for $x \in X$, $\xi(x) = E_x = p^{-1}(x)$ is the *fibre* of ξ over x. By a *Banach family* we mean a fibre set having surjective projection in which each fibre has a given Banach space structure. Thus a Banach bundle is a bundle having open projection and whose underlying fibre set is a Banach family satisfying (3) and (4) of Definition 1.1. If ξ is a fibre set, then $\Pi\xi$ denotes the product of all the fibres of ξ so that if ξ is also a Banach family, then $\Pi\xi$ is a topological vector space. The members of $\Pi\xi$ are called *selections* of ξ, and are the functions $s : X \to E$ such that $p \circ s = id_X$. If ξ is a Banach family over X and $s \in \Pi\xi$, then $|s|$ denotes the function from X to defined by $|s|(x) = \|s(x)\|$, $x \in X$. We let $\Pi_b\xi$ denote the Banach space of *bounded selections* s (selections for which $|s|$ is bounded) with norm given by $\|s\| = \||s|\|_\infty$.

If ξ is a bundle over X, then the space of selections $\Pi\xi$ is a topological product containing the continuous selections as a subspace $\Gamma(\xi)$. We refer to continuous selections as *sections* of ξ. If ξ is a Banach family which is also a bundle over X, then $\Gamma_b(\xi) = (\Pi_b\xi) \quad \Gamma(\xi)$. In case ξ is a Banach bundle $\Gamma_b(\xi)$ is a vector space, which we will see is a closed subspace of $\Pi_b\xi$ and hence a Banach space with norm given by $\|s\| = \||s|\|_\infty$.

By a *Banach algebra bundle* ξ we mean a Banach bundle each of whose fibres is a Banach algebra, and for which the map $E \times_X E \to E$, given by multiplication in each fibre, is continuous. If in addition each fibre of ξ has an isometric involution making it a Banach *-algebra bundle so that the map $E \xrightarrow{*} E$, given by involution in each fibre, is continuous, then we call ξ a *Banach *-algebra bundle* (not to be confused with Banach *-algebraic bundle as defined in [20]). A C^*-*bundle* is a Banach *-algebra bundle each of whose fibres is a C^*-algebra. By a *Hilbert bundle* we mean a Banach bundle each of whose fibres is a Hilbert space.

The notion of an (F)C*-bundle as defined here, coincides with the notion of a C^*-bundle given in [12] and studied in detail in [13], whereas (H)C^*-bundles are used in [3] to represent arbitrary C^*-algebras as section algebras.

Let $\xi = (p,E,X)$ be a Banach bundle. For $S,T \subset E$, let $S + T$ and $S - T$ denote the images under $+$ and $-$, respectively, of $S \times_X T \subset E \times_X E$. For $\varepsilon > 0$ and $\sigma \in \Gamma(\xi)$, we define the ε-tube T_ε by $T_\varepsilon = \{b \in E: \|b\| < \varepsilon\}$, and the ε-*tube about* σ, by $T_\varepsilon(\sigma) = \sigma(X) + T_\varepsilon$. If we refer to a subset S of E satisfying $S - S \subset T_\varepsilon$ as ε-*thin*, then since $T_\varepsilon(\sigma) = \{b \in E: \|b - \sigma(p(b))\| < \varepsilon\}$, it is clear that $T_\varepsilon(\sigma)$

is an open, 2ε-thin set. The nature of the topology of E can now be clarified. If $v \in E$ and $\varepsilon > 0$ then by continuity of subtraction and the norm, each neighborhood of v contains an ε-thin set. Moreover, if $\mathcal{S}$ is any family of neighborhoods of v which contains ε-thin sets for every $\varepsilon > 0$, then it can be shown directly from the definition of a Banach bundle that the family

$$\{S \cap p^{-1}(U): S \in \mathcal{S},\ U \text{ a neighborhood of } p(v)\} \tag{1.1}$$

is a neighborhood base at v in E. For instance, if $\sigma \in \Gamma(\xi)$ and $v = \sigma(x)$, then we can let $\mathcal{S} = \{T_\varepsilon(\sigma) : \varepsilon > 0\}$. If there is a sequence $\sigma_1, \sigma_2, \ldots$ in $\Gamma(\xi)$ with $\sigma_n(x) \to v$, say with $\|\sigma_n(x) - v\| < \frac{1}{n}$, $n = 1,2,\ldots$, then we can let $\mathcal{S} = \{T_{1/n}(\sigma_n): n > 0\}$. In this way it is easy to see that $\Gamma_b(\xi)$ is a Banach space. It should also be noted here that for each $x \in X$, <u>the norm topology on the fibre E_x is the same as the subspace topology which E_x inherits from E.</u>

Suppose now that $Y \subset X$ and ξ is a Banach bundle over X with $\xi = (p,E,X)$. Then $\xi|Y = (p|p^{-1}(Y), p^{-1}(Y), Y)$ is a Banach bundle over Y with operations obtained from ξ by restriction. In general, $|Y$ denotes restriction to Y. Let $r: \Gamma(\xi) \to \Gamma(\xi|Y)$ be the restriction map on sections, and let $r_b: \Gamma_b(\xi) \to \Gamma_b(\xi|Y)$ be the map defined by r on bounded sections. Then r is linear and r_b is a linear contraction. The structure of r_b will allow us to draw several useful conclusions about Banach bundles very quickly, so we examine it more closely. First recall that $U \subset X$ is a <u>halo</u> for $Y \subset X$ if there is a <u>halo function</u> for the pair (U,Y), by which we mean a continuous function $f: X \to [0,1]$ with $f|Y = 1$ and $\operatorname{supp} f \subset U$. Let $\mathcal{H}$ denote the set of all halo functions for (X,Y). Now suppose that $s \in \Gamma(\xi)$, $r(s) \in \Gamma_b(\xi|Y)$ and

$r(s) \neq 0$. If $|s|$ is continuous, we can obtain a halo function $f_0 \in H$ simply by setting

$$f_0 = \frac{\|r(s)\|}{|s| \vee \|r(s)\|},$$

where $\vee$ denotes the pointwise supremum operation on real-valued functions (and $\|r(s)\|$ is treated as a constant function). Notice that $f_0 s$ is a bounded section with $\|f_0 s\| \leq \|r(s)\|$, and $r_b(f_0 s) = r(s)$. Of course, $\|fs\| \geq \|r(s)\|$ for any halo function $f \in H$ for which fs is bounded, so $\|f_0 s\| = \|r(s)\|$. If we cannot assume $|s|$ is continuous, then in any case $|s|$ is upper semicontinuous, and so for any $\varepsilon > 0$ the set

$$U_\varepsilon = \{x \in X : |s|(x) < \|r(s)\| + \varepsilon\}$$

is an open neighborhood of Y. If we suppose that there is a halo function f_ε for (U_ε, Y), then $f_\varepsilon s$ is bounded, $\|f_\varepsilon s\| \leq \|r(s)\| + \varepsilon$, and $r_b(f_\varepsilon s) = r(s)$. Thus if either (1) ξ is an (F) Banach bundle or (2) every open neighborhood of Y in X contains a halo, then (3) $\operatorname{Im} r_b = \operatorname{Im} r \cap \Gamma_b(\xi|Y)$, and (4) if $s \in \Gamma(\xi)$ and $r(s) \in \Gamma_b(\xi|Y)$, then

$$\|r(s)\| = \inf\{\|fs\| : f \in H\} = \operatorname{dist}(0, Hs).$$

In general, for $s \in \Gamma(\xi)$ and $S \subset \Gamma(\xi)$ we have an extension of the norm and distance given by $\|s\| = \||s|\|_\infty \in \mathbb{R}^+ \cup \{\infty\}$ and

$$\operatorname{dist}(s,S) = \inf\{\|s - s'\| : s' \in S\} \in \mathbb{R}^+ \cup \{\infty\}.$$

Now let I be the ideal of functions in $C_b(X)$ which vanish on Y, let $M = \Gamma(\xi)$ and let $K = \text{Ker } r$. Notice that $(1 - H) \subset I$ and $(1 - H)s \subset Is \subset IM \subset K$, so that we have

$$\text{dist}(0, Hs) = \text{dist}(s, (1 - H)s) \geq \text{dist}(s, Is) \geq \text{dist}(s, IM) \geq \text{dist}(s, K)$$
$$= \|s + K\| .$$

Thus if (1) or (2) above holds, then

$$\|s + K\| \leq \text{dist}(s, IM) \leq \|r(s)\|$$

whenever $s \in \Gamma(\xi)$ and $r(s) \in \Gamma_b(\xi|Y)$.

If we begin with $s \in M_b = \Gamma_b(\xi)$, and if $K_b = \text{Ker } r_b$, then as $Is \subset IM_b \subset K_b$, we have

$$\|r_b(s)\| \geq \text{dist}(s, Is) \geq \text{dist}(s, IM_b) \geq \|s + K_b\| .$$

Let $\overline{r}_b : M_b/K_b \to \Gamma_b(\xi|Y)$ be the map induced by r_b. As r_b is a linear contraction, so is $\overline{r}_b$. We now have from the preceding inequalities that

$$\|\overline{r}_b(s + K_b)\| = \|r_b(s)\| \geq \text{dist}(s, IM_b) \geq \|s + K_b\| \geq \|\overline{r}_b(s + K_b)\| ,$$

and hence $\overline{r}_b$ is isometric. Moreover, as all the preceding inequalities are now seen to be equalities, we have $\|s + K_b\| = \text{dist}(s, IM_b)$ for every $s \in M_b$. It follows immediately that if $s \in K_b$, then $\text{dist}(s, IM_b) = 0$, and hence $s \in IM_b$. This show that $K_b = IM_b$. As $\Gamma_b(\xi)$ is a Banach space and $\overline{r}_b$ is isometric, it follows that the image of $\overline{r}_b$ is closed in $\Gamma_b(\xi|Y)$. But $\overline{r}_b$ and r_b have the same image. Hence r_b has closed image in $\Gamma_b(\xi|Y)$.

We summarize the main results of the preceding discussion in the following proposition

PROPOSITION 1.2. *Let* ξ *be a Banach bundle over* X , *and let* $r_b: \Gamma_b(\xi) \to \Gamma_b(\xi|Y)$ *be the restriction map, where* $Y \subset X$. *Also let* $K_b = \text{Ker } r_b$ *and let* I *be the (closed) ideal of functions in* $C_b(X)$ *which vanish on* Y. *Assume that either* (1) ξ *is an* (F) *Banach bundle, or* (2) *every neighborhood of* Y *in* X *contains a halo. Then,*

(3) r_b *induces a linear isometry* $\overline{r}_b: \Gamma_b(\xi)/K_b \to \Gamma_b(\xi|Y)$;

(4) $K_b = I\Gamma_b(\xi)$;

(5) *if* $s \in \Gamma(\xi)$ *and* $s|Y \in \Gamma_b(\xi|Y)$, *then there is a bounded section* $s' \in \Gamma_b(\xi)$ *such that* $s'|Y = s|Y$;

(6) r_b *has closed image in* $\Gamma_b(\xi|Y)$; *and,*

(7) *if* $s \in \Gamma_b(\xi)$, *then* $\|r_b(s)\| = \inf\{\|fs\|: f \in H\}$, *where* H *is the set of all halo functions for* (X,Y). □

The preceding proof of 1.2(3) for the case $Y = \{x\}$, $x \in X$, first appeared in [9] for (F)Banach bundles, and later in [22] for (H)Banach bundles.

Suppose now that either ξ is an (F)Banach bundle or X is completely regular. For each $x \in X$, let $ev_x: \Gamma_b(\xi) \to \xi(x)$ be the evaluation map, let $K_x = \text{Ker } ev_x$, and let I_x be the ideal of functions in $C_b(X)$ vanishing at x. Let

$$\overline{ev}_x: \Gamma_b(\xi)/K_x \to \xi(x)$$

be the map induced by ev_x. Then by Proposition 1.2 $\overline{ev}_x$ is isometric, ev_x has closed image, $K_x = I_x \Gamma_b(\xi)$, and if $v \in \xi(x)$ is in the image of some section of ξ , then it is in the image of some bounded section.

If ξ is a fibred set over X and $S \subset \Pi\xi$, then we set $S(x) = \{s(x): s \in S\}$ for $x \in X$, and we call S full for ξ if $S(x) = \xi(x)$ for each $x \in X$. If ξ is a Banach family we say $S \subset \Pi\xi$ is total for ξ if $S(x)$ spans a dense linear subspace of $\xi(x)$ for each $x \in X$. If ξ is a bundle and if $\Gamma(\xi)$ is full for ξ , then we say ξ is a full bundle.

Suppose now that ξ is a Banach bundle over X and that either X is completely regular or ξ is an (F) Banach bundle. By Proposition 1.2, if ξ is full, then $\Gamma_b(\xi)$ is full for ξ and for each $x \in X$

$$\overline{ev}_x : \Gamma_b(\xi)/K_x \cong \xi(x) \ , \quad K_x = I_x \, \Gamma_b(\xi) \ , \tag{1.2}$$
$$\|s(x)\| = \inf\{\|fs\| : f \in H_x\} \ ,$$

where H_x is the set of halo functions for $(X,\{x\})$. Moreover, by Proposition 1.2, if there is $S \subset \Gamma(\xi)$ total for ξ , then ξ is full and hence $\Gamma_b(\xi)$ is full for ξ. The utility of (1.2) will become clear in the next chapter, as it shows us how to construct the fibres of a Banach bundle which represents a given Banach space by sections.

The basic properties of (F) Banach bundles, as developed for instance in [19], hold also for (H) Banach bundles with only minor modifications in proofs and hypotheses - the case of Proposition 1.2 being rather typical. Thus for (F) Banach bundles the basic properties hold without assuming any separation properties on the spaces involved, whereas for (H) Banach bundles the base space should be assumed at least uniformizable (completely regular) to assure that there are enough continuous functions available. The result of Douady and dal Soglio Herault [20] on the existence of sections for (F) Banach bundles over paracompact Hausdorff base spaces is equally valid for the more general (H) Banach bundles, as the extensive account in [26] shows. Since the base spaces which appear in our applications are usually compact Hausdorff spaces, <u>we assume henceforth that all Banach bundles are full.</u>

As a final application of Proposition 1.2, suppose that ξ is a Banach bundle over the paracompact Hausdorff space X and suppose Y is a closed subset of X. Let $s \in \Gamma_b(\xi|Y)$, $\varepsilon > 0$. Using the fact that ξ is full and that the norm is upper semi-continuous, we can find a partition of unity (f_α) on X and a family of sections (s_α) in $\Gamma_b(\xi)$ such that $\|s_\alpha(x) - s(x)\| < \varepsilon$ whenever $x \in Y \cap \operatorname{supp} f_\alpha$. We set

$$t = \sum_\alpha f_\alpha s_\alpha \in \Gamma_b(\xi) .$$

Then $\|(t|Y) - s\| < \varepsilon$, because for each $x \in Y$, $t(x)$ is a convex combination of points in the ε-ball in $\xi(x)$ centered at $s(x)$. This shows that the restriction map $\Gamma_b(\xi) \to \Gamma_b(\xi|Y)$ has dense range and hence by Proposition 1.2(6) this restriction map is surjective. It follows that each bounded section of $\xi|Y$ extends to a bounded section of ξ. Moreover, if $s \in \Gamma(\xi|Y)$, then the preceding argument still shows that for $\varepsilon > 0$ there is $t \in \Gamma(\xi)$ with $\|(t|Y) - s\| < \varepsilon$. But then there is $t' \in \Gamma_b(\xi)$ with $t'|Y = (t|Y) - s$ and hence $s = (t - t')|Y$. It follows that every section of $\xi|Y$ extends to a section of ξ.

The following proposition appears in [19]. Its proof is straight forward but fairly tedious, so we omit it.

PROPOSITION 1.3. _Let_ $\xi = (p,E,X)$ _be a Banach family and let_ Γ_0 _be a vector subspace of_ $\Pi\xi$ _such that_ (τ is τ^+ or τ_u)

(1) Γ_0 _is total for_ ξ ;

(2) _for each_ $s \in S$, $|s|$ _is_ τ_-continuous._

Then E _has a unique topology so that_ ξ _is a_ (τ) _Banach bundle with_ $\Gamma_0 \subset \Gamma(\xi)$. _Moreover, if each fibre is a Banach algebra (resp., a Banach*-algebra, resp., a C*-algebra) and if_ Γ_0 _is a subalgebra (respectively, a_ $*$_-subalgebra) of_ $\Pi\xi$, _then_ ξ _is a_ (τ) _Banach algebra bundle (resp. a_ (τ) _Banach_ $*$_-algebra bundle, resp., a_ (τ) _C*-bundle)._ □

Proposition 1.3 can be technically generalized in various ways and the details appear in [26] and [27]. Of course, in view of the characterization of the topology on a Banach bundle given in (1.1), the uniqueness assertion of Proposition 1.3 is clear, since the family of all local tubes $T_\varepsilon(s) \cap p^{-1}(U)$ for $s \in \Gamma_0$, $\varepsilon > 0$ and U open in X forms a basis for the topology on E. Thus the proof consists of showing that the hypotheses guarantee that these local tubes do indeed form a basis for a topology and then verifying (1), (3) and (4) in Definition 1.1. Also notice that Proposition 1.3 gives the correspondence between Banach bundles and uniform fields of Banach spaces, as well as the correspondence between (F) Banach bundles and continuous fields of Banach spaces.

A _fibre set map_ $\phi : \xi \longrightarrow \xi'$ is simply a commutative square

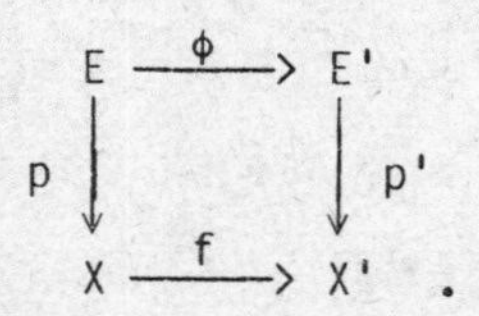

Thus the map on total sets carries fibres into fibres, and for each $x \in X$, we can let $\phi_x : \xi(x) \to \xi'(f(x))$ be the map defined by restricting ϕ. We say that the fibre set map ϕ given by the preceding diagram _covers_ the (continuous) map $f : X \to X'$ or is _over_ f. We say that ϕ is _over_ X to mean that ϕ is over id_X, the identity map on X. If ξ and ξ' are bundles and ϕ is continuous, then $\phi : \xi \to \xi'$ is called a _bundle map covering_ f. If ξ and ξ' are Banach families and ϕ_x is a bounded linear map for each $x \in X$, then $\phi : \xi \to \xi'$ is called a _Banach family map_. If ξ and ξ' are Banach bundles (respectively, C^*-bundles, etc.) and if $\phi : \xi \to \xi'$ is a bundle map which is linear (resp. a *-homomorphism, etc.) on each fibre of ξ, then ϕ is called a _Banach bundle map_ (resp., a _C^*-bundle map_, etc.). Obviously a Banach bundle map is a Banach family map. If $\phi : \xi \to \xi'$ over X is a fibre set map, then we define $\phi_* : \Pi\xi \to \Pi\xi'$ by

$$\phi_*(s) = \phi \circ s , \quad s \in \Pi\xi$$

and note that if ϕ is a bundle map, then ϕ_* carries $\Gamma(\xi)$ into $\Gamma(\xi')$. On the other hand, if $\phi : \xi \to \xi'$ over X is a Banach family map, then $\phi_* : \Pi\xi \to \Pi\xi'$ is linear. Thus, if ξ and ξ' are Banach bundles over X and $\phi : \xi \to \xi'$ is a Banach bundle map over X, then ϕ_* defines a linear map $\phi_* : \Gamma(\xi) \to \Gamma(\xi')$.

If M and N are Banach spaces, then $L(M,N)$ denotes the Banach space of bounded linear maps from M to N. If ξ and ζ are Banach families over X, then $L(\xi,\zeta)$ denotes the Banach family over X whose fibre over $x \in X$ is $L(\xi(x),\zeta(x))$. Let $Hom_X(\xi,\zeta)$ be the vector space of Banach bundle maps of ξ to ζ over X. Then we have

$\phi_x \in L(\xi(x),\zeta(x))$ for $x \in X$ and $\phi : \xi \to \zeta$ a Banach bundle map over X, so we can view

$$\mathrm{Hom}_X(\xi,\zeta) \subset \Pi L(\xi,\zeta) .$$

We let $\mathrm{Hom}_X^b(\xi,\zeta)$ denote the vector subspace of bounded Banach bundle maps of ξ to ζ over X, so

$$\mathrm{Hom}_X^b(\xi,\zeta) \subset \Pi_b L(\xi,\zeta) .$$

We can also view $\Pi L(\xi,\zeta)$ as the vector space of Banach family maps of ξ to ζ over X, if ξ and ζ are any Banach families over X. Thus $\Pi_b L(\xi,\zeta)$ is the Banach space of bounded Banach family maps from ξ to ζ over X with norm $\|\phi\| = \||\phi|\|_\infty$, where now $|\phi|(x) = \|\phi_x\|$. If $\phi : \xi \to \zeta$ is a Banach family map over X and if ϕ is bounded, then for any $s \in \Pi_b\xi$ we have $|\phi_*(s)| \leq |\phi||s|$ and hence

$$|\phi_*(s)| \leq \|\phi\| \; |s| . \tag{1.3}$$

If follows that $\phi_* : \Pi_b\xi \to \Pi_b\zeta$ is a bounded linear map and

$$\|\phi_*\| \leq \|\phi\| . \tag{1.4}$$

Likewise, if ϕ and ψ are composable Banach family maps over X, then $|\phi\circ\psi| \leq |\phi||\psi|$ and hence

$$\|\phi\circ\psi\| \leq \|\phi\|\|\psi\| . \tag{1.5}$$

Of course (1.3), (1.4), and (1.5) hold for Banach bundle maps over X and (1.4) also holds for $\phi_* : \Gamma_b(\xi) \to \Gamma_b(\zeta)$.

If ξ and ζ are Hilbert bundles we say $T \in \mathrm{Hom}_X(\xi,\zeta)$ *has an adjoint* if there is $T^* \in \mathrm{Hom}_X(\zeta,\xi)$ with $T^*(x) = T(x)^*$, for each $x \in X$, or equivalently, that for each $s \in \Gamma_b(\xi)$ and each $t \in \Gamma_b(\zeta)$ we have

$$\langle Ts|t\rangle = \langle s|T^*t\rangle.$$

Obviously T^* is unique and $|T| = |T^*|$. Let $\mathrm{Hom}_X^*(\xi,\zeta)$ be the vector subspace of $\mathrm{Hom}_X^b(\xi,\zeta)$ consisting of those bounded bundle maps having adjoints. Thus $\mathrm{Hom}_X^*(\xi,\xi)$ is a $*$-subalgebra of the C*-algebra $\Pi_b L(\xi,\xi)$. If ξ is a fibre set, say $\xi = (p,E,X)$, and if $Y \subset X$, then $\xi|Y$ is a fibre set and there is an inclusion map $\xi|Y \subset \xi$ which is a fibre set map over the inclusion map $Y \subset X$. If ξ is, respectively, a Banach family, a bundle, or a Banach bundle, then so is $\xi|Y$, and the inclusion $\xi|Y \subset \xi$ is the appropriate type of map. Moreover, in each of these cases, if $\phi : \xi \to \zeta$ is a map over X, there is a unique map $\phi|Y : \xi|Y \to \zeta|Y$ over Y such that the diagram

$$\begin{array}{ccc} \xi & \xrightarrow{\ \phi\ } & \zeta \\ \cup\big\uparrow & & \cup\big\uparrow \\ \xi|Y & \xrightarrow{\ \phi|Y\ } & \xi|Y \end{array}$$

commutes. The mapping ϕ is said to have a property locally if $\phi|U$ has the property for each member U of an open covering of X.

PROPOSITION 1.4. *Let* ξ *and* ζ *be Banach bundles over* X , *and let* $\phi : \xi \to \zeta$ *be a Banach family map over* X. *Then* ϕ *is a Banach bundle map if and only if*

(1) $|\phi| : X \to R^+$ *is locally bounded; and*

(2) *there is a set* S , $S \subset \Gamma_b(\xi)$, *which is total for* ξ , *such that* $\phi_*(S) \subset \Gamma(\zeta)$.

Moreover, $\mathrm{Hom}_X^b(\xi,\zeta)$ *is closed in* $\Pi_b L(\xi,\zeta)$, *and is therefore a Banach space*. *If* ξ *and* ζ *are Hilbert bundles, then* $\mathrm{Hom}_X^*(\xi,\zeta)$ *is closed in* $\mathrm{Hom}_X^b(\xi,\zeta)$ *and thus* $\mathrm{Hom}_X^*(\xi,\xi)$ *is a* C*-*subalgebra of* $\Pi_b L(\xi,\xi)$.

Proof. If $\phi : \xi \to \zeta$ is a Banach bundle map, then (2) holds for $S = \Gamma_b(\xi)$. If $x \in X$, using (1.1) we see there is $\varepsilon > 0$ and an open neighborhood U of x such that $\phi(T_\varepsilon \cap p^{-1}(U)) \subset T_1'$, where T_ε is the ε-tube in ξ and T_1' is the 1-tube in ξ. It follows that $\|\phi(y)\| < \varepsilon^{-1}$ for $y \in U$, so that (2) holds. Conversely, if we assume (1) and (2), then as continuity is a local question and as scalar multiplication is continuous on a Banach bundle, we may as well assume that $\|\phi\| \leq 1$, and hence that $\|\phi_*\| \leq 1$, by (1.4). Let M be the closed linear span of S. Then by Proposition 1.2, M is full for ξ , and as ϕ_* is a continuous linear map with $\phi_*(S) \subset \Gamma_b(\zeta)$, it follows that $\phi_*(M) \subset \Gamma_b(\zeta)$, since $\Gamma_b(\zeta)$ is closed in $\Pi_b\zeta$. If $x \in X$ and $v \in \xi(x)$, then we can choose $s \in M$ with $s(x) = v$. Let $t = \phi_*(s) \in \Gamma_b(\zeta)$. Then for $\varepsilon > 0$, $\phi(T_\varepsilon(s)) \subset T_\varepsilon'(t)$, as $|\phi| \leq 1$, where $T_\varepsilon(s)$ is the ε-tube about s in ξ and $T'_\varepsilon(t)$ is the ε-tube about t in ζ. Hence, for each open neighborhood U of x, $\varepsilon > 0$,

$$\phi(T_\varepsilon(s) \cap p^{-1}(U)) \subset T'_\varepsilon(t) \cap q^{-1}(U) ,$$

where p is the projection of ξ and q that of ζ. Continuity of ϕ now follows from the characterization of the topology of a Banach bundle given by (1.1).

If $\phi \in \Pi_b L(\xi,\zeta)$ is in the closure of $\mathrm{Hom}_X^b(\xi,\zeta)$ and if $s \in \Gamma_b(\xi)$, then for $\varepsilon > 0$, and $\psi \in \mathrm{Hom}_X^b(\xi,\zeta)$ with $\|\phi - \psi\| < \varepsilon$ we have $\|\phi_*(s) - \psi_*(s)\| \leq \varepsilon\|s\|$ and thus $\phi_*(s) \in \Gamma_b(\zeta)$ as $\Gamma_b(\zeta)$ is closed in $\Pi_b\zeta$. Thus (1) and (2) hold for ϕ with $S = \Gamma_b(\zeta)$ so $\phi \in \mathrm{Hom}_X^b(\xi,\zeta)$. □

We say ξ is a <u>subfibre set of</u> ζ <u>over</u> X if ξ and ζ both have base X and $\xi(x) \subset \zeta(x)$ for each $x \in X$; there is then the obvious inclusion fibre set map $i: \xi \subset \zeta$ over X. If ξ and ζ are bundles and the total space of ξ has the topology inherited from that of ζ, then we say ξ is a <u>subbundle of</u> ζ <u>over</u> X, in which case $i: \xi \subset \zeta$ is a bundle map over X. Likewise, if $i: \xi \subset \zeta$ is also a Banach family map, then we say ξ is a <u>Banach subfamily of</u> ζ <u>over</u> X, and if ξ is also a subbundle of ζ over X with ξ and ζ Banach bundles, then we say ξ is a <u>Banach subbundle of</u> ζ <u>over</u> X, in which case $i: \xi \subset \zeta$ is a Banach bundle map over X. Banach algebra subbundles, C*-subbundles, etc., are defined similarly.

If ξ is a Banach subfamily of the Banach bundle ζ over X, then ξ itself is obviously a subbundle of ζ and will be a Banach subbundle of ζ if (1) of Definition 1.1 is true for ξ. In particular, this is always the case if $\Gamma(\zeta) \cap \Pi\xi$ is full for ξ because any bundle having a full set of sections can easily be shown to have an open projection map. In fact, by Proposition 1.3, it is enough to find that $\Gamma(\zeta) \cap \Pi\xi$ is total for ξ. In particular, if M is any vector subspace of $\Gamma(\zeta)$, then there is a unique Banach subbundle ξ of ζ over X with $\xi(x) = \overline{M(x)}$ for each $x \in X$. Moreover, if ζ is a C^*-bundle and M is a *-subalgebra of $\Gamma(\zeta)$, then ξ is a C^*-subbundle.

Let $\phi: \xi \longrightarrow \zeta$ be a Banach bundle map over X. Let M be the image of $\phi_*: \Gamma_b(\xi) \longrightarrow \Gamma_b(\zeta)$. Then there is a unique Banach subbundle ξ' of ζ over X such that for each $x \in X$, $\xi'(x) = \overline{M(x)} \subset \zeta(x)$. Obviously, $\phi: \xi \longrightarrow \xi'$, or more precisely, ϕ factors through the inclusion $\xi' \subset \zeta$, and ξ' is the smallest Banach subbundle of ζ with this property, so we put $\xi' = \mathrm{Im}\,\phi$, and call this <u>the image of</u> ϕ. On the other hand, $\phi(\xi) \subset \zeta$ is the subbundle with $M(x)$ as fibre over $x \in X$, so $\phi(\xi) \subset \mathrm{Im}\,\phi$ and if $\phi(x)$ is isometric for each $x \in X$, then $\phi(\xi) = \mathrm{Im}\,\phi$. The fact that Banach bundle maps have images contrasts sharply with the situation for ordinary vector bundles.

We say that the Banach family map $\psi: \xi \longrightarrow \zeta$, covering $f: X \longrightarrow Y$, is <u>strong</u> provided that for each $x \in X$,

$$\psi(x): \xi(x) \cong \zeta(f(x)), \qquad (1.6)$$

meaning $\psi(x)$ is an isometric isomorphism onto $\zeta(f(x))$, according to our notation. Any $\phi\colon \xi \longrightarrow \zeta$ is called <u>cartesian</u> provided its commutative square is a pullback. If $\phi\colon \xi \longrightarrow \zeta$ covers f, then ϕ is cartesian if and only if for any $\phi'\colon \xi' \longrightarrow \zeta$ covering f there is a unique $\theta\colon \xi' \longrightarrow \xi$ over X such that $\phi' = \phi \circ \theta$. Given a map $f\colon X \longrightarrow Y$ and ζ over Y, we can use the standard pullback construction to produce the <u>induced map</u> $\psi_f(\zeta)\colon f^*(\zeta) \longrightarrow \zeta$ covering f, and if ζ is a Banach family (resp., Banach bundle, C^*-bundle, etc.), then $f^*(\zeta)$ has a unique Banach family (resp., Banach bundle, C^*-bundle, etc.) structure such that $\psi_f(\zeta)$ is strong.

If ξ and ζ are Banach bundles or families, we write $\phi\colon \xi \cong \zeta$ over X and call ϕ an <u>isomorphism</u> (over X) if $\phi\colon \xi \longrightarrow \zeta$ has an inverse and ϕ is strong and covers id_X. Equivalently, ϕ is an isomorphism if ϕ has an inverse and the inequalities

$$\|\phi\| \leq 1 \quad \text{and} \quad \|\phi^{-1}\| \leq 1$$

hold (from which actual equality follows). For C^*-bundles and C^*-bundle maps, we note that if $\psi\colon \xi \longrightarrow \zeta$ is a C^*-bundle map covering $f\colon X \longrightarrow Y$, then for ψ to be strong it is sufficient that $\psi(x)$ be bijective for each $x \in X$. If $\phi\colon \xi \longrightarrow \zeta$ is an invertible Banach bundle or C^*-bundle (etc.) map, then we call it a <u>weak isomorphism</u>. Thus a weak C^*-bundle isomorphism over X is a C^*-bundle isomorphism.

We write $\phi\colon \xi < \zeta$ to mean that $\phi = i \circ \psi$ with $\psi\colon \xi \cong \xi'$ over X and $i\colon \xi' \subset \zeta$ over X. Then for $s \in \Pi\xi$ we have $|\phi_*(s)| = |s|$ and hence $\|\phi_*(s)\| = \|s\|$, so that ϕ_* is a linear isometry. If M and N

are Banach spaces, we write $T: M < N$ to mean that T is a linear isometry. Thus, if $\phi: \xi < \zeta$, then $\phi_*: \Pi_b\xi < \Pi_b\zeta_b$ and if ξ and ζ are Banach bundles, then also $\phi_*: \Gamma_b(\xi) < \Pi_b(\zeta)$. After the next proposition we shall see that if $\phi_*: \Gamma_b(\xi) < \Gamma_b(\zeta)$, then necessarily $\phi: \xi < \zeta$ if either ξ and ζ are (F) Banach bundles or the base space is completely regular.

PROPOSITION 1.5. *If* $\theta: \xi_1 \to \xi_2$ *is a strong Banach bundle map over* X, *then* $\theta: \xi_1 \cong \xi_2$. *If* $\phi: \xi \to \zeta$ *is a strong Banach bundle map, then* ϕ *is cartesian, and if* $\phi': \xi' \to \zeta$ *and* ϕ *cover the same map* $f: X \to Y$, *and if* $\theta: \xi' \to \zeta$ *over* X *is the factorization of* ϕ' *through* ϕ, *then* ϕ' *is strong if and only if* $\theta: \xi' \cong \xi$ *over* X.

Proof. We notice that $\theta(\xi_1) = \xi_2$, so if $M = \mathrm{Im}(\theta_*) \subset \Gamma_b(\xi_2)$, then M is full for ξ_2. Moreover, there is a Banach family map $\psi: \xi_2 \to \xi_1$ over X with $\psi(x) = \theta(x)^{-1}$, for each $x \in X$, and we see $\psi_*(M) \subset \Gamma_b(\xi_1)$, hence by Proposition 1.4, ψ is a Banach bundle map. Thus $\psi = \theta^{-1}$, $\|\psi\| = 1 = \|\theta\|$, and so $\theta: \xi_1 \cong \xi_2$ over X. Let $\phi: \xi \to \zeta$ be a strong Banach bundle map covering $f: X \to Y$. Now $\psi_f = \psi_f(\zeta): f^*(\zeta) \to \zeta$ is cartesian and covers f, so there is a unique Banach bundle map $\theta': \xi \to f^*(\zeta)$ over X such that $\phi = \psi_f \circ \theta'$. Now, ϕ and ψ_f are both strong, hence θ' is also, because $\phi(x) = \psi_f(x) \circ \theta'(x)$ for each $x \in X$. Thus, $\theta': \xi \cong f^*(\zeta)$ over X. Now, if $\phi': \xi' \to \zeta$ also covers $f: X \to Y$, and if $\theta'': \xi' \to f^*(\zeta)$ over X such that $\phi' = \psi_f \circ \theta''$, then with

$\theta = (\theta')^{-1}\circ\theta''$ we have $\phi' = \phi\circ\theta$, and the uniqueness of θ is equally clear. Likewise, if ϕ' is also strong then so is θ and hence $\theta: \xi' \cong \xi$ over X. □

If $\phi: \xi \to \zeta$ is a Banach bundle map over X such that $\phi(x)$ is isometric for each $x \in X$, then $\phi(\xi) = \mathrm{Im}\,\phi \subset \zeta$ and ϕ defines an isomorphism of ξ on $\mathrm{Im}\,\phi$ over X, by Proposition 1.5, and hence $\phi: \xi < \zeta$. Thus $\phi: \xi < \zeta$ if and only if $\phi(x)$ is isometric for each $x \in X$. If ξ and ζ are (F) Banach bundles or X is completely regular and $\phi_*: \Gamma_b(\xi) < \Gamma_b(\zeta)$, then for $x \in X$ and $v \in \xi(x)$ choose $s \in \Gamma_b(\xi)$ with $s(x) = v$ and put $t = \phi_*(s)$. Then $t(x) = \phi(v)$, and if H_x is the set of halo functions for $(X,\{x\})$, then by (1.2),

$$\|v\| = \inf_{f\in H_x} \|fs\| = \inf_{f\in H_x} \|\phi_*(fs)\| = \inf_{f\in H_x} \|ft\| = \|\phi(v)\| .$$

Thus $\phi(x)$ is isometric for each $x \in X$ and hence $\phi: \xi < \zeta$. If $\phi_*: \Gamma_b(\xi) \cong \Gamma_b(\zeta)$, then from the surjectivity of ϕ_* and the fullness of ζ we can conclude that $\phi(\xi) = \zeta$ and hence ϕ is strong. But then $\phi: \xi \cong \zeta$ over X, by Proposition 1.5. We state these facts as a proposition for future reference.

PROPOSITION 1.6. *Let* $\phi: \xi \to \zeta$ *be a Banach bundle map over* X. *Assume either* X *is completely regular or both* ξ *and* ζ *are* (F)*Banach bundles. Then*

(1) $\phi: \xi < \zeta$ *if and only if* $\phi_*: \Gamma_b(\xi) < \Gamma_b(\zeta)$; *and*

(2) $\phi: \xi \cong \zeta$ *if and only if* $\phi_*: \Gamma_b(\xi) \cong \Gamma_b(\zeta)$. □

The importance of Proposition 1.5 to the theory of Banach bundles cannot be over-emphasized. It is not enough to have existence of pullbacks for a workable bundle theory. One must also have an effective characterization of cartesian bundle maps, and this is precisely what Proposition 1.5 provides for the case of Banach bundle maps. Moreover, using Proposition 1.5, all the factorizations and functors typical to bundle theories can be applied. Thus f^* is a functor from the category of Banach bundles and Banach bundle maps over Y to that over X when $f: X \to Y$. If ϕ is a Banach bundle map over Y, then $f^*(\phi)$ is a Banach bundle map over X. If $\phi: \xi_1 \to \xi_2$, then $f^*(\phi)$ is unique so that $\phi \circ \psi_f(\xi_1) = \psi_f(\xi_2) \circ f^*(\phi)$ and $f^*(\phi): f^*(\xi_1) \to f^*(\xi_2)$ over X.

In particular, one might attempt to extend the theory of [13] to include (H) C^*-bundles. However, any straightforward attempt is doomed to failure because the analogue of Theorem 3.5 of [11] can fail very badly for (H) C^*-bundles. In fact we will see that there are many Banach bundles ξ over X for which $\{x \in X: \xi(x) = 0\}$ is an open subset of X.

Let B be a Banach space (respectively, a C^*-algebra). The simplest example of a Banach (resp. C^*-) bundle over X is the trivial or product bundle $\varepsilon(X;B) = (\pi_1, X \times B, X)$ where $\pi_1: X \times B \to X$ is first factor projection. If $\xi \cong \varepsilon(X;B)$ over X, then ξ is called trivial over X with fibre B. Of course any trivial Banach bundle is an (F)Banach bundle, and hence a locally trivial Banach bundle is also an (F)Banach bundle.

If B is a Banach space or a C^*-algebra, and if $\mathrm{Aut}^s B$ is its group of isometric automorphisms with the strong operator topology, then

$\mathrm{Aut}^s B$ is a topological group and the evaluation

$$ev: \mathrm{Aut}^s B \times B \longrightarrow B$$

is continuous. Because of this, if ζ is a Banach bundle or a C^*-bundle over X which is locally trivial with fibre B, then ζ is a fibre bundle with group $\mathrm{Aut}^s B$ and fibre B.

If ξ is a trivial bundle over X with finite dimensional fibre F, and if ζ is any Banach bundle over X, then it is easy to see that $\mathrm{Hom}_X(\xi,\zeta)$ is full for $L(\xi,\zeta)$. Moreover, using the compactness of the unit ball of F, it is easy to see that $|\phi|$ is upper semicontinuous for each $\phi \in \mathrm{Hom}_X(\xi,\zeta)$, just using the method of [11, Section 1]. Thus by Proposition 1.3, $L(\xi,\zeta)$ is a Banach bundle in a unique way so that $\mathrm{Hom}_X(\xi,\zeta) \subset \Gamma(L(\xi,\zeta))$. Moreover, using Proposition 1.4 it is easy to show that $\mathrm{Hom}_X(\xi,\zeta) = \Gamma(L(\xi,\zeta))$. If ξ is locally trivial with finite dimensional fibre F, and if X is completely regular, then again $\mathrm{Hom}_X(\xi,\zeta)$ is full for $L(\xi,\zeta)$ and $|\phi|$ is upper semicontinuous for each $\phi \in \mathrm{Hom}_X(\xi,\zeta)$. Therefore again by Proposition 1.3, $L(\xi,\zeta)$ is a Banach bundle with $\mathrm{Hom}_X(\xi,\zeta) \subset \Gamma(L(\xi,\zeta))$, and since continuity of a bundle map is only a local question, $\mathrm{Hom}_X(\xi,\zeta) = \Gamma(L(\xi,\zeta))$. The development in [11, Section 1] given for (F) Banach bundles actually works for (H) Banach bundles with obvious modifications. In particular, if ζ is also an (F) Banach bundle, then so is $L(\xi,\zeta)$, and if ζ is locally trivial with fibre B, then $L(\xi,\zeta)$ is locally trivial with fibre $L(F,B)$ and is a fibre bundle with group

$$(\mathrm{Aut}^s B \times \mathrm{Aut}\ F)/S^1 = \mathrm{Aut}^s B \times_{S^1} \mathrm{Aut}\ F$$

where we regard the circle S^1 as a subgroup of $\mathrm{Aut}^s B \times \mathrm{Aut}\ F$ via scalar multiples of $(\mathrm{id}_B, \mathrm{id}_F)$ so that S^1 is diagonally embedded in $\mathrm{Aut}^s B \times \mathrm{Aut}\ F$. The main point to observe is that the natural action of $\mathrm{Aut}^s B \times \mathrm{Aut}\ F$ on $L(F,B)$ is easily seen to be continuous using the rule of exponential correspondence together with the fact that F is locally compact so that the norm and compact open topologies agree on $L(F,B)$. If ζ is any Banach bundle over X, then $U(\xi,\zeta)$ denotes the subbundle of $L(\xi,\zeta)$ consisting of isometric linear maps, and hence by our remarks following Proposition 1.5, $\phi \in \Gamma(U(\xi,\zeta))$ if and only if $\phi: \xi < \zeta$, so ϕ embeds ξ as a Banach subbundle of ζ. Moreover, if ζ is again locally trivial with fibre B, then $U(\xi,\zeta)$ is also a fibre bundle with fibre $U(F,B)$ and the same group as $L(\xi,\zeta)$, namely

$$\mathrm{Aut}^s B \times_{S^1} \mathrm{Aut}\ F .$$

More general than locally trivial Banach or C^*-bundles are those obtained from pasting via cocycles of the type discussed in [13, Section 6]. In particular, [13, Theorem 6.1] holds for (H) Banach bundles and (H) C^*-bundles and is actually parallel to the construction in [27]. In particular, if ξ is a Banach or C^*-bundle over the open set $U \subset X$, with X completely regular, then there is a unique Banach bundle over X with $\zeta|U = \xi$ and $\zeta|X\backslash A$ the trivial bundle with zero fibre, and this can be seen directly from Proposition 1.3. In case U is a locally compact Hausdorff space and X is the one point compactification of U, then obviously restriction gives an isometric isomorphism $r: \Gamma(\zeta) \cong \Gamma_0(\xi)$, where $\Gamma_0(\xi)$ denotes the Banach space (respectively, C^*-algebra) of sections vanishing at infinity. In particular, we shall see in the next chapter that this construction is useful for constructing bundles of multipliers. Notice that $r(fs) = r(f)\, r(s)$ for any $f \in C(X)$ and $s \in \Gamma(\zeta)$, and $r(C(X)) \subset C_b(U)$ is the set of functions in $C(U)$ having a limit at infinity. In general, if ξ is a Banach bundle over X, then we denote

$$\operatorname{supp} \xi = \{x \in X : \xi(x) \neq 0\}$$

and call this the support of ξ.

We conclude this chapter with some interesting examples of Banach bundles and C*-bundles.

EXAMPLE 1.7. If $\xi = (p,E,X)$ is any fibred set over the space X and if $\Gamma \subset \Pi\xi$, then giving E the topology coinduced by the family of maps $s : X \to E$ for $s \in \Gamma$, we obtain a topology on E making all selections in Γ continuous. Since $p \circ s = id_X$ is continuous for every $s \in \Gamma$, it follows $p : E \to X$ is continuous. Moreover, since any full bundle has an open projection, if Γ is full for ξ, then p is open. Suppose now that ξ is a Banach family over X, and suppose that $\Gamma \subset \Pi_b\xi$ is a total vector subspace, so $\overline{\Gamma(x)} = \xi(x)$ for each $x \in X$. The preceding construction will give us a bundle structure on ξ over X so that $\Gamma \subset \Gamma_b(\xi)$, but ξ will generally not be a Banach bundle. We have to change the topology of X. Let X be given the weakest topology containing the original topology such that $|s|$ is upper-semicontinuous for each $s \in \Gamma$. With the new topology on X, then ξ has a unique Banach bundle structure such that $\Gamma \subset \Gamma_b(\xi)$, by Proposition 1.3. We can write $\xi[\Gamma]$ for this Banach bundle to denote its dependence on Γ. If ξ is a C*-family and Γ is a *-subalgebra of $\Pi_b(\xi)$ which is total for ξ, then $\xi[\Gamma]$ is a C*-bundle with $\Gamma \subset \Gamma_b(\xi)$. We note that $\xi(\Gamma)$ is a Banach bundle analog of the uniform field given by [3, Theorem I, page 23]. We could also use the weakest topology containing the topology of X so that $|s|$ is continuous for each $s \in \Gamma$, and thereby obtain an (F)Banach bundle.

EXAMPLE 1.8. Let A be a C*-algebra and $X = \text{Prim } A$, the primitive ideal space of A in its hull-kernel topology. Let ξ_A^0 be the C*-family over X whose fibre over $P \in X$ is given by $\xi_A^0(P) = A/P$. For each $a \in A$, let $\hat{a} \in \Pi\xi_A^0$ be the selection defined by $\hat{a}(P) = a + P \in \xi_A^0(P)$, for $P \in X$. Since $\|\hat{a}\| = \|a\|$ is well known (Dixmier [4, 2.7.3]), it follows that $\hat{} : A \to \Pi_b\xi_A^0$ is an injective *-homomorphism. Since $\hat{}(A)$ is actually full for ξ_A^0, we get a C*-bundle

$$\xi_A^{01} = \xi_A^0[\hat{}(A)]$$

so that $\hat{} : A \to \Gamma_b(\xi_A^{01})$ is an injective *-homomorphism. This is the C*-bundle analog of the <u>canonical field</u> of A from [3, 1.14, page 54]. The problem here is that X no longer has the hull-kernel topology and $\hat{}$ is generally not surjective. However, if Prim A is Hausdorff in the hull-kernel topology, then by [4, 3.3.9] we know that $|\hat{a}|$ is continuous for each $a \in A$ and thus ξ_A^{01} is a (F)C*-bundle and we arrive at the representation first obtained by Fell in [18]. In this case $\hat{} : A \cong \Gamma_0(\xi_A^{01})$, as shown in [18], and as will follow from Proposition 2.18, of our next chapter.

EXAMPLE 1.9. Let A be a C*-algebra and begin with the C*-family ξ_A^0 of Example 1.8. For each $P \in \text{Prim } A$, let

$$P_{\#} = \{a \in P : |\hat{a}| \text{ is continuous at } P\}.$$

Then $P_{\#}$ is easily checked to be a closed two-sided ideal of A and obviously $P_{\#} \subset P$. Let $\xi_A^{\vee}$ be the C*-family over $X = \text{Prim } A$ whose fibre over P is $\xi_A^{\vee}(P) = A/P_{\#}$. For each $a \in A$ define $\check{a} \in \Pi_b \xi_A^{\vee}$ by

$$\check{a}(P) = a + P_{\#} \in \xi_A^{\vee}(P).$$

Then $\vee : A \to \Pi_b \xi_A^{\vee}$ is a *-homomorphism and $A^{\vee} = \vee(A)$ is full for $\xi_A^{\vee}$. Moreover, our choice of $P_{\#}$ guarantees that $|\check{a}|$ is upper semicontinuous for each $a \in A$. Another application of Proposition 1.3 now tells us that $\xi_A^{\vee}$ has a unique C*-bundle structure so that $A^{\vee} \subset \Gamma_b(\xi_A^{\vee})$. Since $P_{\#} \subset P$ for each $P \in \text{Prim } A$, it follows that $\vee : A \to \Gamma_b(\xi_A^{\vee})$ is an injective *-homomorphism. The C*-bundle $\xi_A^{\vee}$ was first defined as a uniform field of C*-algebras and studied in [3]. Of course, if Prim A is Hausdorff, then ξ_A^{01} and $\xi_A^{\vee}$ coincide so that $A^{\vee} = \Gamma_0(\xi_A^{\vee})$ in this case. However, there are known cases [32] where $A^{\vee} = \Gamma_b(\xi_A^{\vee})$ when Prim A is compact but not Hausdorff. In general, because $P_{\#} \subset P$ always, there is a unique C*-family map

$$\psi : \xi_A^{\vee} \to \xi_A^0$$

covering the identity and so that $\psi \circ \check{a} = \hat{a}$, for each $a \in A$.

EXAMPLE 1.10. Again, let A be a C*-algebra and let $X = S(A)$ be the state space of A as defined in [4]. Thus $S(A) \subset A^d$ is a subspace of the dual space of A having the weak*-topology. For each $p \in S(A)$, let $\pi_p : A \to \mathcal{B}(H_p)$ be the *-representation of A given by the GNS-construction applied to p. Let $A_1 = A + \mathbb{C}1 \subset M(A)$, where $M(A)$ denotes the multiplier algebra of A. Then $\pi_p : A_1 \to \mathcal{B}(H_p)$. Let $v_p \in H_p$ be the cyclic vector given by the GNS-construction, so that $p(a) = \langle \pi_p(a)v_p | v_p \rangle$ for every $a \in A_1$. Then we can define H_A to be the Banach family over $S(A)$ whose fibre over $p \in S(A)$ is just $H_A(p) = H_p$. For $a \in A$ define $s(a) \in \Pi H_A$ by setting $s(a)(p) = \pi_p(a) v_p$. Thus $s(1)(p) = v_p$, for each $p \in S(A)$. If $a \in A$, then

$$
\begin{aligned}
|s(a)|^2(p) &= \langle \pi_p(a) v_p | \pi_p(a) v_p \rangle \\
&= \langle \pi_p(a^*a) v_p | v_p \rangle = p(a^*a)
\end{aligned}
$$

and as $S(A)$ has the weak*-topology, it follows that $|s(a)|$ is continuous. Also, $\overline{[s(A)](p)} = \overline{\pi_p(A) v_p} = H_p$ for each $p \in S(A)$. Applying Proposition 1.3 we obtain a unique (F)Banach bundle structure on H_A such that $s : A \to \Gamma_b(H_A)$. As each fibre of H_A is a Hilbert space, H_A is an (F)Hilbert bundle and of course by polarization, $\langle | \rangle : H_A \oplus H_A \to \mathbb{C}$ is continuous. If $a \in A$, we can define a Banach family map $\psi(a) : H_A \to H_A$ over $S(A)$ by $\psi(a)w = \pi_p(a)w$ for $w \in H_p$. Since $\psi(a) \circ s(b) = s(ab)$

whenever $a,b \in A$ and as $s(A)$ is total for H_A, it follows from Proposition 1.4 that

$$\psi(a) : H_A \to H_A$$

is a Hilbert bundle map over $S(A)$, and of course it has an adjoint, $\psi(a^*)$. Thus

$$\psi : A \to \mathrm{Hom}^*_{S(A)}(H_A, H_A)$$

is an injective *-homomorphism of C*-algebras.

EXAMPLE 1.11. Let A be a C*-algebra and let $\mathcal{I}(A)$ denote the set of all closed two-sided ideals of A. Let $n : \mathcal{I}(A) \to {}^A$ be the map $[n(I)](a) = \|a+I\|$. The image of n is easily seen to be a closed subset of

$$\prod_{a \in A} [0, \|a\|] \subset \mathbb{R}^A \text{ (product topology)}$$

and hence $n(\mathcal{I}(A))$ is compact. As n is injective, it induces a topology on $\mathcal{I}(A)$ making it a compact Hausdorff space [4, 3.9.2]. Let $\xi_A^{(n)}$ denote the C*-family over $\mathcal{I}(A)$ whose fibre over I is the quotient A/I. Define a *-homomorphism $\overline{n} : A \to \Pi_b \xi_A^{(n)}$ by $\overline{n}(a)(I) = a+I$. Then $\overline{n}(A)$ is a full C*-subalgebra of $\Pi_b \xi_A^{(n)}$. If $a \in A$ we have $|\overline{n}(a)|(I) = n(I)(a)$ So $|\overline{n}(a)| : \mathcal{I}(A) \to \mathbb{R}^A$ is continuous. Therefore, by Proposition 1.3 we can give $\xi_A^{(n)}$ a unique (F)C*-bundle structure so that

$$\overline{n}(A) \subset \Gamma_b(\xi_A^{(n)}).$$

Now Prim $A \subset \mathcal{I}(A)$ as sets and the closure of Prim A inside $\mathcal{I}(A)$, denoted H(Prim(A)), is a compact Hausdorff space. The restriction $\xi_A^{(n)}|H(\mathrm{Prim}(A))$ was first studied by Fell in [18]. As the composition

$$A \xrightarrow{\quad \overline{n} \quad} \Gamma(\xi_A^{(n)}) \xrightarrow{\text{restrict}} \Gamma(\xi_A^{(n)}|H(\mathrm{Prim}\ A))$$

is an injective *-homomorphism, we see that any C*-algebra can be represented as a full C*-subalgebra of an algebra of sections of an (F)C*-bundle over a compact space. Unfortunately, it is possible that $0 \in H(\mathrm{Prim}\ A)$, so that A itself occurs as one of the fibres. In general, if Prim A is not Hausdorff, there will be sections of $\xi_A^{(n)}|H(\mathrm{Prim}\ A)$ which are not restrictions of sections in $\overline{n}(A)$.

2 Banach modules

Let A be an algebra over $\mathbb{K} = \mathbb{R}$ or $\mathbb{C}$. We say M is an A-module (resp., an A-algebra) to mean M is an algebra (left) A-module (resp., A-algebra) which is unital if A has an identity.

Let A be a Banach algebra. We say M is a Banach A-module (resp., a Banach A-algebra) if M is a Banach space (resp., a Banach algebra) which is an A-module (resp., an A-algebra) and such that

$$\|am\| \leq \|a\|\|m\|$$

for $a \in A$, $m \in M$.

Suppose X is a space and ξ is a Banach bundle over X. The continuity of scalar multiplication guarantees that $\Gamma(\xi)$ is a $C(X)$-module and clearly $\Gamma_b(\xi)$ is a Banach $C_b(X)$-module. Moreover, from Proposition 1.2 we see that if $M = \Gamma_b(\xi)$, then for each $x \in X$,

$$\xi(x) \cong M/K_x$$

where $K_x = \text{Ker } ev_x$, and $K_x = I_x M$, where I_x, as before, is the ideal in $C_b(X)$ of functions vanishing at x. Thus, the fibres of ξ are determined by the $C_b(X)$-module structure and we can see clearly now how to construct the fibres for a prospective Banach bundle when we begin with an arbitrary Banach $C_b(X)$-module.

Let M be a Banach $C_b(X)$-module. We seek a Banach bundle ξ over X such that M and $\Gamma_b(\xi)$ are isomorphic as Banach $C_b(X)$-modules; in other words $M \cong \Gamma_b(\xi)$. Of course this is too much to ask without certain additional requirements on M and X; requirements which will naturally appear in what follows. Let K_x be the closed linear span of $I_x M$, and set $E_x = M/K_x$ for each $x \in X$. By our previous remarks and Proposition 1.2 it is clear that E_x is the logical candidate for $\xi(x)$. Certainly E_x is a Banach $C_b(X)$-module because K_x is a $C_b(X)$-submodule of M. In fact, it is readily shown that $fv = f(x)v$ for $f \in C_b(X)$ and $v \in E_x$, since $I_x M \subset K_x$. Let $\xi = (p,E,X)$ be the Banach family with $\xi(x) = E_x$ for each $x \in X$. Thus E is the disjoint union of the sets E_x for $x \in X$ and $p : E \to X$ is the projection map with $p^{-1}(x) = E_x$. For $s \in M$ let $\tilde{s} \in \Pi\xi$ be defined by $\tilde{s}(x) = s + K_x$, $x \in X$, and let $\tilde{M} = \{\tilde{s} : s \in M\} \subset \Pi\xi$. Thus $\tilde{M}$ is a $C_b(X)$-submodule of $\Pi\xi$ and $\sim : M \to \Pi\xi$ is a $C_b(X)$-module homomorphism. Obviously, $\tilde{M}(x) = \xi(x)$ for each $x \in X$ so we are nearly in a position to apply Proposition 1.3 with $\Gamma_0 = \tilde{M}$.

The final obstacle to application of Proposition 1.3 is neatly and cleverly solved by the following proposition due to Januario Varela [42,3.2]. For each $x \in X$, let N_x denote the family of neighborhoods of x in X.

PROPOSITION 2.1. _Let_ x _be a fixed point of_ X _and for each_ $V \in N_x$, _suppose_ $f_V : X \to [0,1]$ _is continuous,_ $f_V^{-1}(1) \in N_x$, _and_ f_V _vanishes off_ V. _Then for each_ $s \in M$,

$$\|\tilde{s}(x)\| = \inf\{\|f_V s\| : V \in N_x\}.$$

Proof. Given $\varepsilon > 0$, let $t \in K_x$ be chosen so that $\|s + t\| < \|\tilde{s}(x)\| + \varepsilon$. Choose $h_1,\ldots,h_n$ in I_x and $t_1,\ldots,t_n$ in M satisfying

$$\|t - (h_1 t_1 + \ldots + h_n t_n)\| < \varepsilon$$

and set $t_0 = h_1 t_1 + \ldots + h_n t_n$. Now with b as an upper bound for $\{\|t_1\|, \ldots, \|t_n\|\}$, choose $V \in N_x$ so that $\|h_i|V\| < \varepsilon/nb$ for $1 \leq i \leq n$. It follows that $\|f_V t_0\| < \varepsilon$. Consequently we have

$$\begin{aligned}\|\tilde{s}(x)\| + \varepsilon > \|s + t\| &\geq \|f_V s + f_V t\| \geq \\ &\|f_V s + f_V t_0\| - \|f_V t_0 - f_V t\| \geq \\ &\|f_V s\| - \|f_V t_0\| - \|t_0 - t\| \geq \\ &\|f_V s\| - 2\varepsilon,\end{aligned}$$

and thus $\|\tilde{s}(x)\| \geq \inf\{\|f_V s\| : V \in N_x\}$.

On the other hand, if $V \in N_x$, then $f_V s = s + (f_V - 1)s$, and $f_V - 1$ belongs to I_x, so that $f_V s$ belongs to $s + K_x$. It follows that $\|\tilde{s}(x)\| \leq \|f_V s\|$ for any $V \in N_x$, and we have the desired equality. □

COROLLARY 2.2. _If X is completely regular, then for each_ $s \in M$, _the function $|\tilde{s}|$ is upper semicontinuous on_ X.

Now suppose that X is completely regular. Combining Proposition 1.3 and Corollary 2.2, we see that E has a unique topology making $\xi = (p,E,X)$ an (H)-Banach bundle with $\tilde{M} \subset \Gamma_b(\xi)$, and that

$$\sim : M \to \Gamma_b(\xi)$$

is a $C_b(X)$-module homomorphism of M into $\Gamma_b(\xi)$ satisfying $\|\tilde{s}\| \leq \|s\|$. We can think of $\tilde{s}$ as a sort of generalized Gelfand transform of s. Of course the Banach bundle ξ we have constructed depends on the Banach $C_b(X)$-module M, and moreover, a given Banach space M can often be considered as a Banach $C_b(X)$-module for several different spaces X. We denote these dependencies by by writing $\xi = {}_X\xi_M$ and $\sim\, = G_X(M)$ whenever M is a Banach $C_b(X)$-module, remaining free to drop any subscripts whose values are clear from context. Now, notice that if M is a Banach $C_b(X)$-module, then $\tilde{M}$ is a $C_b(X)$-submodule of $\Gamma_b(\xi_M)$ and obviously $\tilde{M}$ is full for ξ_M.

PROPOSITION 2.3. *Suppose* X *is compact Hausdorff,* ζ *is a Banach bundle over* X, *and* M' *is a* $C(X)$-*submodule of* $\Gamma(\zeta)$ *such that* M' *is full for* ζ. *Then* M' *is dense in* $\Gamma(\zeta)$. *Hence, if* M *is a Banach* $C(X)$-*module, then* $\tilde{M}$ *is dense in* $\Gamma(\xi_M)$.

Proof. Simply apply the proof of [19, 1.7] which is equally valid for (H)Banach bundles. Thus, if $\varepsilon > 0$ and if $s \in \Gamma_b(\zeta)$, then there is a finite partition of unity $f_1,\ldots,f_n$ on X and $s_i \in M'$, $1 \leq i \leq n$ so that for x in the support of f_i we have $\|s_i(x) - s(x)\| < \varepsilon$. Now setting $s_0 = f_1 s_1 + \ldots + f_n s_n$ we have $\|s_0 - s\| < \varepsilon$. □

We note that $K_M = \cap\{K_x : x \in X\}$ is the kernel of the transform $\sim : M \to \Gamma_b(\xi_M)$ so that if $K_M = 0$, then $\sim$ is an injective, contractive $C_b(X)$-module homomorphism into $\Gamma_b(\xi_M)$ which has dense range if X is also compact Hausdorff.

Now let B be a C*-algebra with identity, 1, and let A be a C*-subalgebra of $Z(B)$ with $1 \in A$. Let $X = \hat{A}$ and identify A with $C(X)$. Then

B is a Banach A-module, and as A is contained in the center of B, K_x is a two sided ideal of B for each $x \in X$. Thus each fibre of ξ_B is in fact a C*-algebra and $\tilde{B}$ is a *-subalgebra of $\Pi\{\xi_B(x) : x \in X\}$. It follows from Proposition 1.3 that ξ_B is an (H)C*-bundle and thus $\Gamma_b(\xi_B)$ is a C*-algebra. But also, $\sim : B \to \Gamma_b(\xi_B)$ is a *-homomorphism, so by Proposition 2.3 we have $\tilde{B} = \Gamma_b(\xi_B)$. If π is an irreducible representation of B, then π must carry $Z(B)$ onto scalar multiples of the identity and hence $\pi|A = x1$ where $x \in \hat{A}$ and 1 is the identity operator. This shows $I_x \subset \operatorname{Ker} \pi$. But with $I_x \subset P \in \operatorname{Prim}(B)$ we have $K_x \subset P$ and therefore each irreducible representation of B factors through $ev_x \circ \sim : B \to \xi_B(x)$, for some $x \in X$. In particular, this means that

$$K_B = \cap\{K_x : x \in X\} \subset \cap\{P : P \in \operatorname{Prim}(B)\} = 0,$$

and therefore $B \cong \Gamma_b(\xi_B)$ under $\sim$.

If B has no identity, then we simply pass to $M(B)$, the multiplier algebra of B. If A is a C*-subalgebra of $Z(M(B))$ containing $1 \in M(B)$, and if $X = \hat{A}$, we immediately obtain a unique C*-subbundle ξ_B of $\xi_{M(B)}$ such that $\tilde{B}(x) = \xi_B(x) \subset \xi_{M(B)}(x)$ for $x \in X$. By Proposition 2.3 we have that $\sim : M(B) \cong \Gamma(\xi_{M(B)})$ carries B *-isomorphically onto $\Gamma(\xi_B)$. Since A is an ideal of $M(A)$, every irreducible representation of A extends to an irreducible representation of $M(A)$ and hence factors through $ev_x \circ \sim$ for some $x \in X$. In case $A = Z(M(B))$, this is one of the main results of [3] for C*-algebras.

With the following theorem we summarize the results so far for C*-algebras, results which were first obtained by Varela in [42], and which generalize one of the main results of [3].

THEOREM 2.4. *If B is a C*-algebra and A is a C*-subalgebra of the center of $M(B)$ containing an identity for $M(B)$, and if $X = \hat{A}$, then there is a *-isomorphism*

$$\sim : B \cong \Gamma(\xi_B)$$

of B onto the C-algebra of all sections of ξ_B. Moreover, every irreducible representation of $\Gamma(\xi_B)$ factors through a point evaluation.* □

The advantage of the preceding development over that used in [3] and [42] is that we have circumvented the need for dealing with the field star topology and subsequently relating it to the hull-kernel topology.

If M is a Banach $C_b(X)$-module we can of course look for conditions on M which might guarantee that $\sim : M \to \Gamma(\xi_M)$ is isometric. This has already been solved by K. H. Hofmann [27]. We say that M is a $C_b(X)$-*convex module* provided M is a Banach $C_b(X)$-module such that for any pair $f,g \in C_b(X)^+$ with $f + g = 1$, and any pair $s,t \in M$, we have

$$\|fs + gt\| \leq \sup\{\|s\|,\|t\|\}.$$

It follows immediately that if $f_1,\ldots,f_n$ is a finite partition of unity for X and if $s_1,\ldots,s_n \in M$, then

$$\|f_1 s_1 + \ldots + f_n s_n\| \leq \sup_i \|s_i\|.$$

It is obvious that $\Gamma_b(\xi)$ is a $C_b(X)$-convex module whenever ξ is a Banach bundle over X. Thus, by Theorem 2.4, each C*-algebra B is $Z(M(B))$-convex.

THEOREM 2.5. <u>If</u> M <u>is a</u> $C(X)$-<u>convex module with</u> X <u>a compact Hausdorff space, then</u>

$$\sim : M \cong \Gamma(\xi_M)$$

<u>is an isometric isomorphism onto</u> $\Gamma(\xi_M)$.

Proof. Suppose $\varepsilon > 0$. Using Proposition 2.1 and the compactness of X we can find, for fixed s in M, a finite open cover $\{U_1,\ldots,U_n\}$ of X and continuous functions $f_i : X \to [0,1]$ with $U_i \subset f_i^{-1}(1)$ and such that $\|f_i s\| < \|\tilde{s}\| + \varepsilon$, $1 \leq i \leq n$. Choose a partition of unity $g_1,\ldots,g_n$ for X so that $\operatorname{supp} g_i \subset U_i$, $1 \leq i \leq n$. Then we have

$$\|s\| = \|\sum_1^n g_i f_i s\| \leq \sup_i \|f_i s\| \leq \|\tilde{s}\| + \varepsilon.$$

Thus $\|s\| \leq \|\tilde{s}\|$, but already $\|\tilde{s}\| \leq \|s\|$, so that $\sim : M \to \Gamma(\xi_M)$ is isometric and an application of Proposition 2.3 is all that is needed. □

The correspondence from Banach $C_b(X)$-modules to Banach bundles over X is really part of a functor which we will need to construct. If M and N are Banach A-modules, then $\operatorname{Hom}_A^b(M,N)$ denotes the Banach subspace of $L(M,N)$ consisting of all bounded A-module homomorphisms. Notice that if A is commutative, then the natural A-module structure on $\operatorname{Hom}_A^b(M,N)$ makes $\operatorname{Hom}_A^b(M,N)$ into a Banach A-module. Now, take $A = C_b(X)$. If ξ and ζ are Banach bundles over X, then $\operatorname{Hom}_X(\xi,\zeta)$ is a $C(X)$-module because scalar multiplication is continuous on a Banach bundle. Moreover, $\operatorname{Hom}_X^b(\xi,\zeta)$ is an A-submodule, and as $|f\phi| = |f||\phi|$ we have $\|f\phi\| \leq \|f\|\|\phi\|$, $f \in A$, $\phi \in \operatorname{Hom}_X^b(\xi,\zeta)$, and hence by Proposition 1.4, $\operatorname{Hom}_X^b(\xi,\zeta)$ is itself a Banach

A-module. If $\phi \in \mathrm{Hom}_X^b(\xi,\zeta)$, put $\Gamma_b(\phi) = \phi_* \in \mathrm{Hom}_A^b(M,N)$, $M = \Gamma_b(\xi)$, $N = \Gamma_b(\zeta)$. Thus, $\Gamma_b : \mathrm{Hom}_X^b(\xi,\zeta) \to \mathrm{Hom}_A(M,N)$ is, by (1.4), a contractive linear map and is therefore a contractive A-module homomorphism. Let $\mathcal{B}(X)$ be the category of Banach bundles and bounded Banach bundle maps over X, let $\mathcal{M}(X)$ be the category of Banach A-modules and bounded A-module homomorphisms, and let $\mathcal{M}_c(X)$ be the full subcategory of $\mathcal{M}(X)$ consisting of A-convex modules. Let $I_X : \mathcal{M}_c(X) \subset \mathcal{M}(X)$ be the inclusion functor. We define $\Gamma_X : \mathcal{B}(X) \to \mathcal{M}_c(X)$ by $\Gamma_X(\xi) = \Gamma_b(\xi)$ and $\Gamma_X(\phi) = \phi_*$, so Γ_X is a functor which we could describe as a contractive A-linear functor.

Suppose X is completely regular. If M is a Banach A-module, let $K_x(M) = \overline{\mathrm{span}\ I_x M}$ for $x \in X$. If $T \in \mathrm{Hom}_A^b(M,N)$, then by continuity of T we have $T(K_x(M)) \subset [T(\mathrm{span}\ I_x M)]^-$, and hence by A-linearity of T, we have $T(K_x(M)) \subset K_x(N)$. It follows that there is for each $x \in X$, a unique continuous linear map $T_x : \xi_M(x) \to \xi_N(x)$ such that

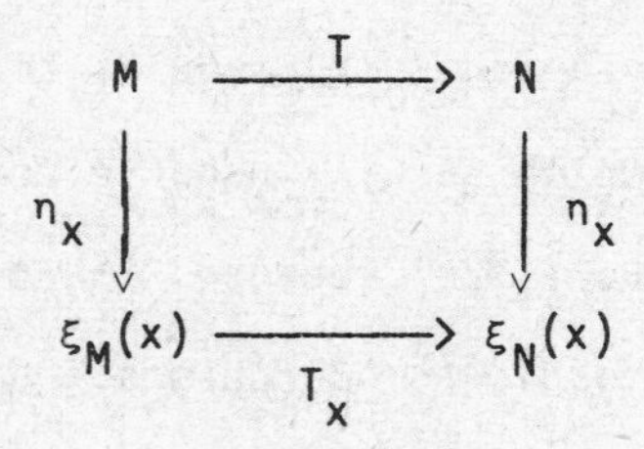

$$\begin{array}{ccc} M & \xrightarrow{T} & N \\ \downarrow \eta_x & & \downarrow \eta_x \\ \xi_M(x) & \xrightarrow{T_x} & \xi_N(x) \end{array}$$

is a commutative diagram, where the vertical maps are just the natural quotient maps. But then $\|T_x\| \leq \|T\|$ for each $x \in X$, and $T_x(\tilde{s}(x)) = \widetilde{T(s)}(x)$ for $s \in M$. Thus there is $\phi : \xi_M \to \xi_N$ over X a Banach family map with $\phi(x) = T_x$, $\phi_*(\tilde{s}) = \widetilde{T(s)}$, $x \in X$, $s \in M$. But then by Proposition 1.4,

$\phi : \xi_M \to \xi_N$ is a Banach bundle map as $\widetilde{M}$ is full for ξ_M, and ϕ is bounded with $\|\phi\| \leq \|T\|$, because $\phi(x) = T_x$, $x \in X$. We put $R_X(T) = \phi$, and set $R_X(M) = \xi_M$. This gives us a functor $R_X : \mathcal{M}(X) \to \mathcal{B}(X)$ which is also a contractive A-linear functor. Moreover, if $G(M) : M \to \Gamma_X R_X(M)$ is the Gelfand transform $\sim : M \to \Gamma_b(\xi_M)$, then we have a commutative diagram

$$\begin{array}{ccc} M & \xrightarrow{G(M)} & \Gamma_X R_X(M) \\ {\scriptstyle T}\downarrow & & \downarrow{\scriptstyle \phi_* = \Gamma_X R_X(T)} \\ N & \xrightarrow{G(N)} & \Gamma_X R_X(N) \end{array}$$

and thus $G : id_{\mathcal{M}(X)} \to I_X \circ \Gamma_X \circ R_X$ is a natural transformation of functors, which is a natural A-linear transformation of functors. On the other hand, if ξ is a Banach bundle over X, then by (1.2), for each $x \in X$, there is a unique $\theta_x : \xi(x) \cong \Gamma_b(\xi)/I_x\Gamma_b(\xi)$ such that

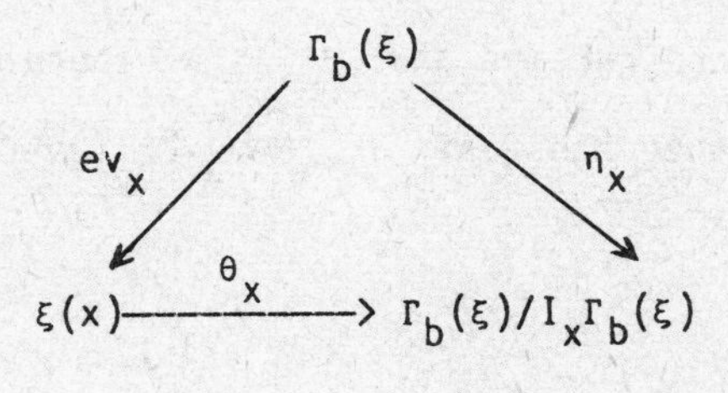

commutes. Putting $M = \Gamma_b(\xi) = \Gamma_X(\xi)$, we have $\xi_M(x) = \Gamma_b(\xi)/I_x\Gamma_b(\xi)$, and thus $\theta_x : \xi(x) \cong \xi_M(x)$ for each $x \in X$. This defines a Banach family map

$\theta : \xi \to \xi_M$ over X which is strong, and if $s \in \Gamma_b(\xi)$, then for $x \in X$,

$$\theta_*(s)\ (x) = \theta_x(s(x)) = \eta_x(s) = \tilde{s}(x)$$

hence $\theta_* = G(M)$ so by Proposition 1.4 and Proposition 1.5, $\theta : \xi \cong \xi_M$ over X. The naturality of θ is easy to check, so that we have

$$\theta : \mathrm{id}_{\mathcal{B}(X)} \cong R_X \circ I_X \circ \Gamma_X \tag{2.1}$$

is an (isometric) isomorphism of functors. Suppose that $\phi : \xi \to \zeta$ over X and that either ξ is an (F)Banach bundle or X is completely regular. If $x \in X$ and $v \in \xi(x)$ with $\|v\| < 1$, then by (1.2) there is $s \in \Gamma_b(\xi)$ with $\|s\| < 1$ and $s(x) = v$. Then

$$\|\phi(v)\| = \|\phi_*(s)\ (x)\| \leq \|\phi_*(s)\| \leq \|\phi_*\|$$

and hence $\|\phi\| \leq \|\phi_*\|$. But $\|\phi_*\| \leq \|\phi\|$ already as Γ_X is contractive, hence $\|\phi\| = \|\phi_*\|$. Thus Γ_X is isometric. In case X is completely regular, this fact is also obvious from (2.1) as then $\|\phi\| = \|R_X \circ I_X \circ \Gamma_X(\phi)\| \leq \|\phi_*\|$, since R_X and I_X are also contractive and θ is a natural Banach bundle isomorphism. On the other hand, if $T \in \mathrm{Hom}_A^b(\Gamma_b(\xi), \Gamma_b(\zeta))$, then using (1.2) and

$$T(I_x\Gamma_b(\xi)) \subset I_xT(\Gamma_b(\xi)) \subset I_x\Gamma_b(\zeta) \subset \mathrm{Ker}\ ev_x,$$

it follows that there is a unique Banach family map $\phi : \xi \to \zeta$ over X such that $\phi_*(s) = T(s)$ for each $s \in \Gamma_b(\xi)$. But then by Proposition 1.4,

$\phi \in \mathrm{Hom}_X^b(\xi,\zeta)$, and $\Gamma_b(\phi) = T$. Thus

$$\Gamma_b : \mathrm{Hom}_A^b(\xi,\zeta) \cong \mathrm{Hom}_A^b(\Gamma_b(\xi), \Gamma_b(\zeta)) \tag{2.2}$$

is an isometric A-module isomorphism. Because a linear contraction with contractive linear inverse is necessarily isometric, (2.2) can be used, together with the fact that Γ_X is a functor, to provide an alternative proof of Proposition 1.6(2). Suppose now that X is a compact Hausdorff space and M and N are Banach A-modules. Further, suppose that N is C(X)-convex. For $T \in \mathrm{Hom}_A(M,N)$ and $\phi = R_X(T)$, we have

$$\begin{array}{ccc} M & \xrightarrow{\;G(M)\;} & \Gamma(R_X(M)) \\ {\scriptstyle T}\big\downarrow & & \big\downarrow{\scriptstyle \phi_*} \\ N & \underset{G(N)}{\xrightarrow{\;\cong\;}} & \Gamma(R_X(N)) \end{array}$$

is a commutative diagram, $G(N)$ being an isometric isomorphism by Proposition 2.5. Then $\|T\| = \|\phi_* G(M)\| \le \|\phi_*\| = \|\phi\|$ by (2.2), and hence $\|T\| = \|R_X(T)\|$, and R_X is also isometric. If $\phi : \xi_M \to \xi_N$ is a Banach bundle map over X, then we can use the preceding diagram to define T, as $G(N)$ is an isomorphism. But then $\Gamma_X \circ R_X(T) = \Gamma_X(\phi)$, because $G(M)$ has dense range by Proposition 2.3, and both $\Gamma_X \circ R_X(T)$ and $\Gamma_X(\phi)$ fit in the diagram. This proves that for X a compact Hausdorff space and N a C(X)-convex module,

$$R_X : \mathrm{Hom}_A^b(M,N) \cong \mathrm{Hom}_X^b(\xi_M,\xi_N) \tag{2.3}$$

is an isometric A-module isomorphism. If $T \in \mathrm{Hom}_A^b(M,N)$ and $R_X(T): \xi_M \cong \xi_N$, and if both M and N are $C(X)$-convex, then in the preceding diagram we also have $G(M) : M \cong \Gamma(R_X(M))$ and $\phi_* = R_X(T)_* : \Gamma(\xi_M) \cong \Gamma(\xi_N)$, hence $T : M \cong N$. Conversely, using (2.3) and the fact that R_X is a functor, we see that if $T : M \cong N$, then $R_X(T) : \xi_M \cong \xi_N$. If $R_X(T) : \xi_M < \xi_N$, then again, the preceding diagram shows that $T : M < N$. Conversely, if $T : M < N$, then the preceding diagram gives $\Gamma_X \circ R_X(T) : \Gamma(\xi_M) < \Gamma(\xi_N)$ and hence by Proposition 1.6, we conclude $R_X(T) : \xi_M < \xi_N$.

The functors R_X and Γ_X can be used to represent Banach A-module valued functors of Banach bundles over X. An example will suffice to make the idea clear. If ξ and ζ are Banach bundles over X, then $\mathrm{Hom}_X^b(\xi,\zeta)$ is a Banach A-module and putting $\mathrm{HOM}_X(\xi,\zeta) = R_X(\mathrm{Hom}_X^b(\xi,\zeta))$, we have that the Gelfand transformation is a contractive A-linear map

$$G : \mathrm{Hom}_X^b(\xi,\zeta) \to \Gamma_b(\mathrm{HOM}_X(\xi,\zeta)),$$

which represents arbitrary bounded Banach bundle maps as continuous sections. However, there does not seem to be any clear connection between $\mathrm{HOM}_X(\xi,\zeta)$ and $L(\xi,\zeta)$, although if ξ is locally trivial with finite dimensional fibre, then it is not hard to see that $\mathrm{HOM}_X(\xi,\zeta) = L(\xi,\zeta)$. Moreover, if $Y \subset X$, then $\mathrm{HOM}_X(\xi,\zeta)|Y$ and $\mathrm{HOM}_Y(\xi|Y,\zeta|Y)$ are not related nicely enough in general to allow the section extension process to give extensions of Banach bundle maps. On the other hand, if ξ is locally trivial with finite dimensional fibre, then the methods of [11, Section 1] show that for any $f : X' \to X$ we have

$$f^*(L(\xi,\zeta)) \cong L(f^*(\xi), f^*(\zeta))$$

over X' so that $L(\xi,\zeta)$ is well behaved in this case.

Suppose now that A is any C*-algebra. By an A-normed module M we mean a Banach space which is an A-module together with a function $|\cdot| : M \to A^+$, where $A^+ = \{a \in A : a \geq 0\}$, satisfying

(1) for $m,n \in M$, $|m + n| \leq |m| + |n|$;

(2) for $a \in A$, $m \in M$, $|am| \leq |a||m|$;

(3) for $m \in M$, $\| |m| \| = \|m\|$.

Of course, for $a \in A$ we have $|a| = (a^*a)^{1/2}$, by definition. Notice that if ξ is an (F)Banach bundle over X, then $\Gamma_b(\xi)$ is a $C_b(X)$-normed module with $|s|(x) = \|s(x)\|$, as already defined. If M is an A-normed module, we have, using (2) and (3),

$$\|am\| = \| |am| \| \leq \| |a||m| \| \leq \| |a| \| \| |m| \| = \|a\| \|m\|,$$

so that M is a Banach A-module.

Suppose now that $A = C_b(X)$ and M is a $C_b(X)$-normed module. For each $x \in X$, let $K_x' = \{m \in M : |m|(x) = 0\}$. If $f \in I_x$ and $m \in M$, then

$$|fm|(x) \leq |f|(x)|m|(x) = |f(x)||m|(x) = 0,$$

and therefore $K_x \subset K_x'$, as K_x' is a closed linear subspace of M. Conversely, suppose that $m \in K_x'$ and $\varepsilon > 0$. Let $U = \{x \in X : |m|(x) < \varepsilon\}$, and let $f = \inf\{(1/\varepsilon)|m|, 1\}$ (pointwise), so that $f : X \to [0,1]$, $f \in I_x$, and $f|(X\setminus U) = 1$. Then

$$|m - fm| = |(1 - f)m| < |1 - f||m| < \varepsilon,$$

and hence $m \in \overline{I_x m} \subset K_x$. This shows that $K_x = K'_x$ for each $x \in X$. Moreover, the seminorm $m \to |m|(x)$ on M induces a norm on $E_x = M/K_x$, so that the natural map $\eta_x : M \to E_x$ has the property that $\|\eta_x(m)\| = |m|(x)$. Let E'_x denote the Banach space completion of E_x in the norm just described, let E' be the disjoint union of the E'_x for $x \in X$, and let $\xi' = (p', E', X)$, where p' is the projection mapping E'_x to x. Then for each $m \in M$, $\tilde{m}$ is a selection of ξ' such that $\|\tilde{m}(x)\| = |m|(x)$. Thus, $\tilde{M}$ is a vector subspace of $\Pi\{E'_x : x \in X\}$ and by Proposition 1.3 there is a unique (F)Banach bundle structure on ξ' such that $\tilde{M} \subset \Gamma_b(\xi')$. Notice that

$$\|m\| = \||m|\| = \||\tilde{m}|\| = \|\tilde{m}\|$$

and hence $\sim$ is isometric. But then, by Proposition 1.2, we can conclude that $E_x = E'_x$ as Banach spaces so that for each $x \in X$ and $m \in M$,

$$|m|(x) = \|m + K_x\| \qquad (2.4)$$

and therefore, $\xi_M = \xi'$. Thus, the $C_b(X)$-normed modules are just those for which the Banach bundle representation gives an (F)Banach bundle when X is completely regular.

A Hilbert Z-module [15, 31, 35, 39] with Z a C*-algebra, is a Banach space M which is a Z-module (unital, if Z has identity), together with a sesquilinear map $M \times M \to Z$, denoted $(m,n) \longmapsto \langle m|n\rangle$ (a sesquilinear map is linear in the first variable and conjugate linear in the second variable) which satisfies

(1) $\langle m|m\rangle \geq 0$ for all $m \in M$;

(2) if $m \in M$ and $\langle m|m\rangle = 0$, then $m = 0$,

(3) $\langle zm|n\rangle = z\langle m|n\rangle$ for all $m,n \in M$, $z \in Z$.

Applying (1) to both $m = a + b$ and $m = ia + b$ and expanding, we find that both $\langle a|b\rangle+\langle b|a\rangle$ and $i\langle a|b\rangle - i\langle b|a\rangle$ are self-adjoint, and from this we conclude that $\langle a|b\rangle = \langle b|a\rangle^*$ in Z. Now suppose $Z = C_b(X)$ for some space X. As the Cauchy-Schwartz inequality holds for positive semi-definite inner products, we can conclude that it holds pointwise, and thus we have $|\langle m|n\rangle| \leq |m||n|$ for all $m,n \in M$, where $|m| = \langle m|m\rangle^{1/2}$ for $m \in M$. It follows that

$$|m + n|^2 = \langle m + n|m + n\rangle = \langle m|m\rangle + 2\mathrm{Re}\langle m|n\rangle + \langle n|n\rangle$$
$$\leq |m|^2 + 2|m||n| + |n|^2 = (|m| + |n|)^2,$$

and hence in view of (3) we conclude that under $|\cdot| : M \to Z^+$, M is also a Z-normed module. Moreover, equation (2.4) allows us to conclude that $M_x = M/K_x$ is a Hilbert space for each $x \in X$. If we understand a <u>Hilbert bundle</u> to be a Banach bundle in which each fibre is a Hilbert space, we see that under our correspondence between Banach modules and Banach bundles, Hilbert modules correspond to (F)Hilbert bundles [10]. We point out that these considerations show that the Z-normed modules and Hilbert Z-modules are all Z-convex

If M is a Banach space we denote its dual space by M^d so $M^d = L(M, \mathbb{K})$, and if $T \in L(M,N)$, then $T^d \in L(N^d,M^d)$ denotes the dual transformation. The bilinear pairing of a Banach space and its dual will simply be denoted $\langle\ ,\ \rangle$. Let A be a Banach algebra. If M is a Banach space, then specifying a Banach A-module structure on M is equivalent to giving a contractive Banach algebra homomorphism of A into $L(M) = L(M,M)$ which must be required to be unital when A has identity. As $T \longmapsto T^d$ is an isometric unital antihomomorphism $L(M) \to L(M^d)$, we see that if M is a left Banach A-module, then M^d is a right Banach A-module with

$$\langle \mu a, m \rangle = \langle \mu, am \rangle$$

for $a \in A$, $\mu \in M^d$, $m \in M$; and consequently, M^{dd} is again a left Banach A-module with

$$\langle am, \mu \rangle = \langle m, \mu a \rangle \tag{2.5}$$

for $m \in M^{dd}$, $a \in A$, $\mu \in M^d$. This Banach A-module structure on M^{dd} can clearly be obtained from that on M directly from the isometric unital homomorphism $T \to T^{dd}$ of $L(M) \to L(M^{dd})$. Thus, M is a Banach A-submodule of M^{dd}.

If M is a Banach A-module, then we denote by $L : A \to L(M)$ the associated Banach algebra homomorphism, so $\|L\| \leq 1$ and $L(a)m = am$, $a \in A$, $m \in M$. If M is a right Banach A-module, we let $R : A^{op} \to L(M)$ be the associated Banach algebra homomorphism, so $R(a)m = ma$, $a \in A$, $m \in M$.

It is useful to note that if M is a right Banach A-module (respectively, a left Banach A-module), then $L(a) \in L(M^d)$ is weak *-continuous (likewise for $R(a)$, respectively). Thus if the A-module structure on M satisfies certain identities, these can sometimes be extended to M^{dd}, as M is weak* dense in M^{dd} and M^{dd} is the dual of a Banach A-module. For example, suppose that B is a C*-algebra and a Banach A-algebra. Recall that B^{dd} is also a C*-algebra under the Arens product (both products coincide on B^{dd} as B is a C*-algebra). In fact, B^{dd} is a von Neumann algebra whose ultraweak topology coincides with the weak* topology on B^{dd}. Thus $L(b) \in L(B^{dd})$ and $R(b) \in L(B^{dd})$ are weak* continuous for $b \in B^{dd}$ or $b \in A$. Now, B^{dd} can be shown also to be a Banach A-algebra by a direct calculation with the Arens product, but that is rather tedious. If $a \in A$ and $b \in B$, the algebra condition $(ab)c = a(bc)$ for $c \in B$ says that $L(ab)$ agrees with $L(a)\,L(b)$ on $B \subset B^{dd}$ and hence by weak* continuity of both maps, we have $L(ab) = L(a) \circ L(b)$, or $(ab)c = a(bc)$ for any $a \in A$, $b \in B$, $c \in B^{dd}$, and this says that for $a \in A$ and $c \in B^{dd}$, the weak* continuous maps $R(c) \circ L(a)$ and $L(a) \circ R(c)$ agree on B so that $R(c) \circ L(a) = L(a) \circ R(c)$, hence $(ab)c = a(bc)$ for any $a \in A$, $b,c \in B^{dd}$, and thus, B^{dd} is a Banach A-algebra.

Suppose now that A is a C*-algebra. We say B is an A-*<u>algebra</u> to mean that B is a C*-algebra which is both a left and right Banach A-algebra such that

(1) $(am)^* = m^*a^*$, $a \in A$, $m \in B$,

(2) $(am)b = a(mb)$, $a,b \in A$, $m \in B$,

(3) $(ka)m = k(am)$, $a \in A$, $k,m \in B$.

As the adjoint operation on B^{dd} is weak* continuous, (1) must also hold for B^{dd}, and (2) says $R(b) \circ L(a)$ and $L(a) \circ R(b)$ agree on B, hence (2) also holds for B^{dd}, and a similar two step argument shows that (3) also holds for B^{dd}, so B^{dd} is also an A-*algebra. Again, we have avoided a tedious calculation with the Arens products. One might be tempted to think that B^{dd} is an A^{dd}-*algebra, but this does not seem to be the case, as there are two possible Arens type "actions" of A^{dd} on B^{dd} which "share" the identities.

We say that B is a <u>central</u> A-*<u>algebra</u> if in addition B satisfies $am = ma$ for $a \in A$, $m \in B$, or equivalently, $(am)^* = a^*m^*$, $a \in A$, $m \in B$. In case A is commutative and the C*-algebra B is a Banach A-algebra, if B satisfies $(am)^* = a^*m^*$, $a \in A$, $m \in B$, then as the left Banach A-algebra structure can be used to make B also a right Banach A-algebra, we see that B is a central A-*algebra.

Suppose now that B is an A-*algebra. Then B^{dd} is also. Let 1_B denote the identity of B^{dd}. We can define $\phi : A \to B^{dd}$ by $\phi(a) = a1_B$. Then $\phi(ab) = (ab)1_B = a(b1_B) = a(1_B(b1_B)) = (a1_B)(b1_B) = \phi(a)\phi(b)$, and as $1_Ba = (1_ba)1_B = a1_B$, $\phi(a^*) = a^*\, 1_B = (1_Ba)^* = (a1_B)^* = \phi(a)^*$, $a,b \in A$. Thus $\phi : A \to B^{dd}$ is a *-homomorphism. If $b \in B^{dd}$ and $a \in A$, then

$$\begin{aligned} \phi(a)b &= (a1_B)b = ab \\ b\phi(a) &= b(a1_B) = b(1_Ba) = ba \end{aligned} \tag{2.6}$$

so that the A-*algebra structure of B^{dd} is completely determined by the *-homomorphism $\phi : A \to B^{dd}$ which is unital if A has identity. Now, recall that we can define the <u>multiplier</u> <u>algebra</u> $M(B)$ of B by

$$M(B) = \{b \in B^{dd} : bB \subset B \subset Bb\} \tag{2.7}$$

so $M(B)$ is a C*-subalgebra of B^{dd}. If $a \in A$, $b \in B$, then $\phi(a)b = ab \in B$ and $b\phi(a) = ba \in B$, so $\phi : A \to M(B)$. Thus the *-homomorphism $\phi : A \to M(B)$ completely specifies the A-*algebra structure on B. Conversely, if we are given a *-homomorphism (unital if A has identity) of A into $M(B)$, say $\psi : A \to M(B)$, then clearly setting $ab = \psi(a)b$ and $ba = b\psi(a)$ for $a \in A$, $b \in B$ makes B into an A-*algebra. Of course, in this case we have $\psi = \phi$, since for $a \in A$,

$$\phi(a) = a1_B = \psi(a)1_B = \psi(a). \quad (2.8)$$

Moreover, clearly B _is a central_ A-*_algebra if and only if_ $\phi : A \to Z(M(B))$. Now, if A, A' are any Banach algebras and M a Banach A-module, then a Banach algebra homomorphism $\phi : A' \to A$ which is contractive (and unital if A' has identity) can always be used to make M into a Banach A'-module by setting $a'm = \phi(a')m$ for $a' \in A'$, $m \in M$. But, _if_ $\phi : A' \to A$ _is a unital_ *-_homomorphism of abelian_ C*-_algebras and if_ M _is an_ A-_convex module_, _then_ M _is also_ A'-_convex_, _since_ $a \in A'$, $a \geq 0$ _implies_ $\phi(a) \in A^+$. _Thus by Theorem_ 4, _and these remarks_, _if_ A _is a commutative_ C*-_algebra with identity and if_ B _is a central_ A-*_algebra_, _then_ B _is_ A-_convex_, _since_ B _is_ $Z(M(B))$_-convex and_ $\phi : A \to Z(M(B))$ _gives the_ A-_module structure on_ B.

Suppose now that $A = C_b(X)$ and B is a central A-*algebra. For $x \in X$, $K_x(B) = [\text{span}(I_xB)]^-$ is a closed two-sided ideal of B, as $I_xB = BI_x$, and hence $\xi_B(x)$ is also a C*-algebra. Also, the Gelfand transformation is a *-homomorphism $G(B) : B \to \Pi_b\xi_b$, and hence its image, $\tilde{B}$ is a *-subalgebra of $\Pi_b\xi_B$, but $\tilde{B} \subset \Gamma_b(\xi_B)$, hence by Proposition 1.3, ξ_B is actually a C*-bundle over X. On the other hand, if ξ is any C*-bundle over a space Y, then $\Gamma_b(\xi)$ is obviously a central $C_b(Y)$-*algebra and if $\alpha : \xi \to \zeta$ over Y is a C*-bundle map, then necessarily $\|\alpha\| \leq 1$, and $\alpha_*:\Gamma_b(\xi) \to \Gamma_b(\zeta)$ is a $C_b(Y)$-*homomorphism. If $\phi : B_1 \to B_2$ is a $C_b(X)$-*homomorphism of central $C_b(X)$-*algebras, then $R_X(\phi)$ is a C*-bundle map, as factorizations of *-homomorphisms through surjective *-homomorphisms always produce *-homomorphisms.

If $\mathcal{C}^*(X)$ denotes the category of central $C_b(X)$-*algebras and $C_b(X)$-*-homomorphisms, and if $\mathcal{C}^*\mathcal{B}(X)$ denotes the category of C*-bundles and C*-bundle maps over X, then we have forgetful functors $F : \mathcal{C}^*(X) \to \mathcal{M}_c(X)$, as each $C_b(X)$-*algebra is $C_b(X)$-convex, and $F: \mathcal{C}^*\mathcal{B}(X) \to \mathcal{B}(X)$, and our natural Banach bundle isomorphism $\theta : id_{B(X)} \cong R_X \circ I_X \circ \Gamma_X$ gives a natural C*-bundle isomorphism

$$\theta_* : id_{C^*B(X)} \cong R_X^* \circ \Gamma_X^* \tag{2.9}$$

such that $\theta_* \circ F = \theta$. Likewise, the Gelfand transform $G : id_{B(X)} \to \Gamma_X \circ R_X \circ I_X$ gives the Gelfand transform

$$G_* : id_{C^*(X)} \to \Gamma_X^* \circ R_X^*$$

and if X is a compact Hausdorff space, by Theorem 2.5, we must have

$$G_* : id_{C^*(X)} \cong \Gamma_X^* \circ R_X^* , \tag{2.10}$$

where $R_X^* : C^*(X) \to C^*B(X)$ and $\Gamma_X^* : C^*B(X) \to C^*(X)$ are the obvious functors so $F \circ R_X^* = R_X \circ F$ and $F \circ \Gamma_X^* = \Gamma_X \circ F$.

In the next chapters we will make extensive use of these results for the case where X is a compact Hausdorff space. In this case, by Proposition 1.4, $\mathrm{Hom}_X^b(\xi,\zeta) = \mathrm{Hom}_X(\xi,\zeta)$ for any Banach bundles ξ and ζ over X, and hence $B(X)$ becomes the category of Banach bundles and Banach bundle maps over X.

For purposes of future reference, we summarize the results we will need in the following theorem.

THEOREM 2.6. Let X be a compact Hausdorff space and $Z = C(X)$. Then

$$\Gamma_X : B(X) \longrightarrow M_c(X)$$

is an equivalence of categories. The reverse equivalence is given by the representation functor

$$R_X^c = R_X \circ I_X : \mathcal{M}_c(X) \to \mathcal{B}(X),$$

and the general Gelfand transformation provides the natural isometric Z-*isomorphism*

$$G : id_{\mathcal{M}_c(X)} \cong \Gamma_X \circ R_X.$$

Also, for $T \in Hom_Z^b(M,N)$ *we have*

(1) $T : M \cong N$ if and only if $R_X(T) : \xi_M \cong \xi_N$,

and

(2) $T : M < N$ if and only if $R_X(T) : \xi_M < \xi_N$.

Moreover, under the equivalence R_X, *the* Z-*normed modules correspond exactly to the* (F)*Banach bundles, the Hilbert* Z-*modules correspond exactly to the* (F)*Hilbert bundles, and central* Z-**algebras and* Z-**homomorphisms correspond exactly to the* C*-*bundles and* C*-*bundle maps over* X. □

Because of Theorem 2.6(2), if X is a compact Hausdorff space and M is a $C(X)$-convex module (note that then every Banach $C(X)$-submodule of M is again $C(X)$-convex), then we can regard $\xi_N \subset \xi_M$ for each Banach $C(X)$-submodule N of M, since if $i : N \subset M$ is the inclusion, we have $R_X(i) : \xi_N < \xi_M$ so we replace ξ_N by $ImR_X(i)$. As R_X is a functor, if $N_1 \subset N_2 \subset M$, then $\xi_{N_1} \subset \xi_{N_2} \subset \xi_M$, so that no trouble with other inclusions results from this modification. Henceforth, we will make this identification without explicit mention. Moreover, if N is a Banach submodule of M a $C(X)$-convex module, then $\tilde{N}(x) = \xi_N(x) \subset \xi_M(x)$, where $\tilde{N}$ is the image of N under $G(M) : M \cong \Gamma(\xi_M)$, and $G(N)$ becomes simply $G(M)|N$. In particular, if ξ is a C*-bundle over X and if J is a closed two-sided ideal of $\Gamma(\xi)$, then as $J^+ \subset J^2$ and $J = \text{span } J^+$, it follows that J is a $C(X)$-submodule of $\Gamma(\xi)$ so by the preceding observation and Proposition 2.3 *the equality* $J = \Gamma(\xi_J)$ *holds, where* $\xi_J \subset \xi$ *is the* C*-*subbundle of* ξ *defined by* $\xi_J(x) = J(x) \subset \xi(x)$ *for* $x \in X$, *a result generalizing a result of J.M.G. Fell* [4, 10.4.2].

Let $\xi = (p,E,X)$ be a Banach bundle over a completely regular Hausdorff space. By (4) of Proposition 1.2 we know that $K_x = I_x\Gamma_b(\xi)$ for each $x \in X$. On the other hand, we know that if x and y are distinct points of X, then $I_x + I_y = C_b(X)$ and hence

$$K_x + K_y = \Gamma_b(\xi).$$

It follows that any non-zero linear map with domain $\Gamma_b(\xi)$ *can factor through at most one point evaluation*. Suppose X is compact and $A = \Gamma(\xi)$ with ξ a C*-bundle. Then by the uniqueness in Theorem 2.6 we can identify ξ_A with ξ. If π is an irreducible representation of A, then π is non-zero so by Theorem 2.4 there is a unique $x \in X$ such that π factors through $ev_x : A \to \xi(x)$. This means we have a map $\hat{p} : \text{Prim } A \to X$ such that for $x \in X$. $\hat{p}^{-1}(x)$ is the set of primitive ideals of A containing K_x. Also, $\hat{A}$ has the topology induced by the surjection $\text{Ker} : \hat{A} \to \text{Prim}(A)$ sending the unitary equivalence class $[\pi]$ to $\text{Ker } \pi$, and consequently $\text{Ker} : \hat{A} \to \text{Prim}(A)$ is a continuous open map so we can equally well view Prim A as having the quotient topology from $\hat{A}$ induced by Ker. For simplicity, let $\hat{p} \circ \text{Ker} : \hat{A} \to X$ also be denoted by $\hat{p}$. Thus $\hat{p} : \hat{A} \to X$ is continuous (respectively, open) if and only if $\hat{p} : \text{Prim } A \to X$ is continuous (respectively, open). Define

$$\hat{\xi} = (\hat{p}, \hat{A}, X)$$

$$\text{Prim}(\xi) = (\hat{p}, \text{Prim}(A), X)$$

Following [4, 3.2.1], if J is a closed two-sided ideal of A, then we identify $\hat{J}$ with the open set $\hat{A}^J \subset \hat{A}$ and Prim J with the open set $\text{Prim}^J(A) \subset \text{Prim}(A)$. Obviously since $ev_x : \Gamma(\xi) \to \xi(x)$ is surjective, we have $\hat{\xi(x)} \subset \hat{A}$ via the identification for quotients given in [4, 3.2.1], and obviously

$$\hat{\xi(x)} = \hat{p}^{-1}(x).$$

It is also clear that for any closed two-sided ideal J of A, since $J = \Gamma(\xi_J)$ we must have

$$\hat{p}(\hat{J}) = \text{supp}(\xi_J) \qquad (*)$$

and in particular,

$$\hat{p}(\hat{A}) = \text{supp } \xi.$$

On the other hand, if J has the property that $\xi_J(x) = \xi(x)$ for each $x \in \text{supp } \xi_J$, then

$$\hat{J} = \hat{p}^{-1}(\text{supp } \xi_J). \qquad (\overset{*}{*})$$

In particular, if U is open in X, then taking J to be the kernel of the restriction map $\Gamma(\xi) \to \Gamma(\xi|X\backslash U)$, we obtain $U = \text{supp}\ \xi_J$ and $\xi_J(x) = \xi(x)$ for each $x \in U$. Thus $\hat{J} = \hat{p}^{-1}(U)$ by $(\overset{*}{*})$ and this shows $\hat{p} : \hat{A} \to X$ is continuous. If ξ is an (F)C*-bundle, then so is any C*-subbundle of ξ. But supports of (F)Banach bundles are always open and thus $\hat{p}$ is an open map by (*). Conversely suppose that p is an open map. Then by (*) for each closed two-sided ideal J of A, $\text{supp}(\xi_J)$ is open. For b a self-adjoint element in a C*-algebra, we denote by b^+ its positive part [4, 1.5.7]. Notice that for an element a in a C*-algebra with identity we have $\|a\| > \varepsilon$ if and only if $(a^*a - \varepsilon^2 1)^+ \neq 0$. Now A is a central C(X)-*algebra so the C(X)-module structure of A is given by a *-homomorphism $C(X) \to ZM(A)$ as we demonstrated in our remarks preceding Theorem 2.6. Thus A is a C(X)-*subalgebra of M(A) and we have $\xi \subset \xi_{M(A)}$. If $a \in A$ we put $b = a^*a - \varepsilon^2 1$. As $ev_x : M(A) \to \xi_{M(A)}(x)$ is a *-homomorphism we have

$$ev_x(b^+) = [ev_x(b)]^+.$$

Let J be the closed linear span of Ab^+A. As A is an ideal of M(A), J will be a closed two-sided ideal of A and

$$\hat{p}(\hat{J}) = \text{supp}\ \xi_J = \{x \in X : (b(x))^+ \neq 0\} = \{x \in X : |a|(x) > \varepsilon\}.$$

Thus $|a|$ is continuous being now both upper and lower semicontinuous. But then by Proposition 1.3, it follows that ξ is an (F)C*-bundle. Thus ξ is an (F)C*-bundle if and only if $\hat{p} : \hat{A} \to X$ is an open map. It is easy to see that if $\alpha : \xi \cong \zeta$ over X is a C*-bundle isomorphism, then it naturally gives rise to $\hat{\alpha} : \hat{\xi} \cong \hat{\zeta}$ over X so that if $\beta : \zeta \to \zeta'$ over X, then $\widehat{\beta\circ\alpha} = \hat{\beta}\circ\hat{\alpha}$. It is also well known that if $\zeta = \varepsilon(X;F)$, then $\hat{\zeta} = \varepsilon(X,\hat{F})$. Thus, if ξ is locally trivial with fibre F, then $\hat{\xi}$ is locally trivial with fibre $\hat{F}$. In fact, if ξ is a fibre bundle with fibre F and group G, then $\hat{\xi}$ is a fibre bundle with fibre $\hat{F}$ and group $\hat{G}$ where $\hat{G} \subset (\hat{F})^{\hat{F}}$ is given the compact-open topology, because $\hat{F}$ is locally compact (but possibly not Hausdorff).

PROPOSITION 2.7. *If ξ is a C*-bundle over the compact Hausdorff space X, then $\hat{p}_\xi : \widehat{\Gamma(\xi)} \to X$ is a continuous map and is open if and only if ξ is an (F)C*-bundle. Moreover, for each $x \in X$, $\hat{p}_\xi^{-1}(x) = \widehat{\xi(x)} \subset \widehat{\Gamma(\xi)}$, so with $\hat{\xi} = (\hat{p}_\xi, \widehat{\Gamma(\xi)}, X)$, $\hat{\xi}$ is a bundle over X with $(\hat{\xi})(x) = \widehat{\xi(x)}$, and if ξ is locally trivial with fibre F, then $\hat{\xi}$ is locally trivial with fibre $\hat{F}$.* □

The results of Proposition 2.7 for (F)C*-bundles were first obtained by R. Y. Lee [34, Theorem 4].

If B is any C*-algebra, then by the Dauns-Hofmann theorem, B is a $C_b(\hat{B})$-*algebra. If $p : \hat{B} \to Y$ is any continuous map into the completely regular space Y, then $p^* : C_b(Y) \to C_b(\hat{B})$ is a *-homomorphism which gives B a $C_b(Y)$-*algebra structure, and thus we obtain a *-homomorphism $G_* : B \to \Gamma_b({}_Y\xi_B)$. In case $Y = X$, our compact Hausdorff space, we have

$G^\star : B \cong \Gamma(\xi_B)$ and after identifying B and $\Gamma(\xi_B)$ via $G^\star$ we find $\hat{p}_\xi = p$.

Suppose that $\xi = (p,E,X)$ is a Banach bundle over an arbitrary space X, and suppose M is a vector subspace of $\Gamma_b(\xi)$ which is full for ξ. Let ζ be a bundle over Y which is simultaneously a Banach family and suppose

$$\Phi : M \longrightarrow \Gamma_b(\zeta) \subset \Pi_b\zeta$$

is a bounded linear map. Assume also that $ev_x|M : M \to \xi(x)$ is a quotient map for each $x \in X$. This last condition is automatic by Proposition 1.2 in case $M = \Gamma_b(\xi)$, and of course it is automatic if ξ is a C*-bundle and M is a C*-subalgebra of $\Gamma_b(\xi)$. For each $x \in X$, let $Y_x(\Phi)$ denote the set of points $y \in Y$ for which $ev_y \circ \Phi$ factors through $ev_x|M : M \to \xi(x)$, and let

$$Y(\Phi) = \bigcup_{x \in X} Y_x(\Phi).$$

Denote by $\xi_\Phi = (p_\Phi, E(\Phi), X)$ the subbundle of the product bundle $\varepsilon(X;Y)$ whose fibre over $x \in X$ is just

$$\xi_\Phi(x) = \{x\} \times Y_x(\Phi).$$

Let $h = h_\Phi : E(\Phi) \to Y$ be the restriction of the second factor projection. Then $h^*(\zeta)$ is a bundle over $E(\Phi)$ which is also a Banach family. Thus $h^* : \Pi\zeta \to \Pi h^*(\zeta)$ carries $\Gamma(\zeta)$ into $\Gamma(h^*(\zeta))$ and defines a linear contraction $\Pi_b\zeta \to \Pi_b h^*(\zeta)$. If $z = (x,y) \in E(\Phi)$, then $p_\Phi(z) = x$, $h(z) = y$, and $ev_y \circ \Phi$ factors through $ev_x|M$. After identifying $[p_\Phi^\star(\xi)](z) = \xi(x)$ and identifying $[h^*(\zeta)](z) = \zeta(y)$, we obtain a factorization of $ev_z \circ h^* \circ \Phi$ through $p_\Phi^* \circ ev_x|M$ which we denote by

$$\Phi_z : [p_\Phi^*(\xi)](z) \longrightarrow [h^*(\zeta)](z).$$

The family of bounded linear maps (Φ_z) indexed by $E(\Phi)$ now defines a Banach family map

$$\tilde{\Phi} : p_\Phi^*(\xi) \longrightarrow h_\Phi^*(\zeta)$$

over $E(\Phi)$ unique such that the diagram

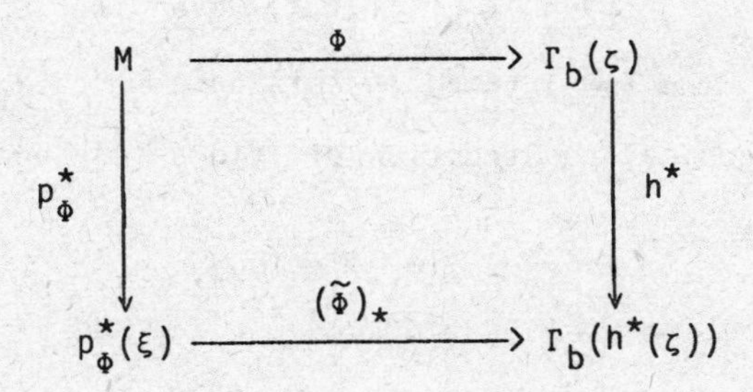

is commutative. Since the Banach bundle map $\psi : p_\Phi^*(\xi) \to \xi$ induced by ξ over p_Φ is itself isometric on fibres and $\psi \circ ev_z \circ p_\Phi^*|M = ev_x|M$ when $x = p_\Phi(z)$, it follows that $ev_z \circ p_\Phi^* : M \to [p_\Phi^*(\xi)](z)$ is also a quotient map for each $z \in E(\Phi)$. Thus $\tilde{\Phi}(z) = \Phi_z$ is bounded by $\|\Phi\|$ for each $z \in E(\Phi)$ so $\tilde{\Phi}$ is a bounded Banach family map. We call $\tilde{\Phi}$ the <u>disintegration</u> of Φ and we call ξ_Φ <u>the base of integration</u> of Φ.

REMARK 2.8. *If* ζ *is a Banach bundle over* Y, *then*

$$\tilde{\Phi} : p_\Phi^*(\xi) \longrightarrow h_\Phi^*(\zeta)$$

is a bounded Banach bundle map over $E(\Phi)$. Indeed, $S = p_\Phi^*(M)$ is full for $p_\Phi^*(\xi)$ and from the commutative diagram we see that $(\tilde{\Phi})_*$ carries S into $\Gamma_b(h^*(\xi))$, so Proposition 1.4 applies to give the result.

REMARK 2.9. *If* ξ *and* ζ *are C*-bundles, if* M *is a C*-subalgebra of* $\Gamma_b(\xi)$, *and if* Φ *is a *-homomorphism, then* $\tilde{\Phi}$ *is a C*-bundle map.*

REMARK 2.10. *If* $f : Y \to X$ *is a continuous map such that* $Y_x(\Phi) = f^{-1}(x)$ *for each* $x \in X$, *then* $\xi_\Phi \cong (f,Y,X)$ *over* X, *so* p_Φ *can be identified with* $f : Y \to X$. Indeed, in this case we note that for $\beta = (f,Y,X)$ the canonical construction of $(id_X)^*(\beta)$ equals ξ_Φ, so

$$\xi_\Phi = (id_X)^*(\beta) \underset{X}{\cong} \beta,$$

and

$$\tilde{\Phi} : f^*(\xi) \to \zeta \quad \text{over} \quad Y$$

is unique so that

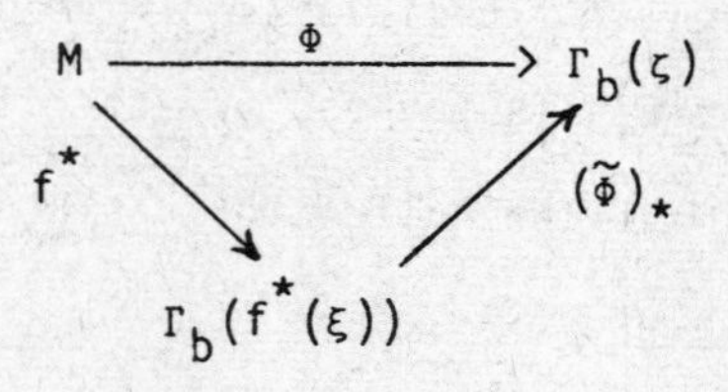

is commutative.

REMARK 2.11. *If* X *is compact, then* $E(\Phi)$ *is closed in* $X \times Y$ *and* $h_\Phi : E(\Phi) \to Y$ *is a closed map. In particular, now*

$$h_\Phi(E(\Phi)) = Y(\Phi)$$

is closed in Y. To see this, first observe that since X is compact the second factor projection $\pi_2 : X \times Y \to Y$ is itself a closed map because every neighborhood of a compact rectangle contains a rectangular neighborhood. Thus, it suffices to show that $E(\Phi)$ is closed in $X \times Y$. Suppose (z_α) is a net in $E(\Phi)$ converging to $z \in X \times Y$ and set $z_\alpha = (x_\alpha, y_\alpha)$ and $z = (x,y)$. Then $p_\Phi(z_\alpha) = x_\alpha$, (x_α) converges to x and (y_α) converges to y. To show $ev_y \circ \Phi$ factors through $ev_x|M : M \to \xi(x)$ we must show that if $m \in M$ and $m(x) = 0$, then $\Phi(m)(y) = 0$. As $|m|$ is upper semi-continuous, $\lim \| m(x_\alpha) \| = 0$ and as $\Phi(m) \in \Gamma(\zeta)$ we have $\lim \Phi(m)(y_\alpha) = \Phi(m)(y)$. But $\tilde{\Phi}$ is a bounded Banach family map with $\| \tilde{\Phi} \| \leq \| \Phi \|$ so

$$\begin{aligned} \| \Phi(m)(y) \| &= \| \lim \Phi(m)(y_\alpha) \| \\ &= \| \lim \tilde{\Phi}(m(x_\alpha)) \| \leq \| \Phi \| \lim \| m(x_\alpha) \| = 0. \end{aligned}$$

EXAMPLE 2.12. Suppose Y is completely regular, M is a Banach subspace of $\Gamma_b(\xi)$, and M has a $C_b(Y)$-module structure. We can take $\zeta = {}_Y\xi_M$ and

$$\Phi = G_Y(M) : M \longrightarrow \Gamma_b(\zeta).$$

Then
$$\tilde{\Phi} : p_\Phi^*(\xi) \longrightarrow h_\Phi(\zeta)$$

is a bounded Banach bundle map. If $f : Y \to X$, if $M = \Gamma_b(\xi)$, and if the $C_b(X)$-module structure of M actually comes from the $C_b(Y)$-module structure via f^*, then for $y \in Y$ and $x = f(y)$ we have

$$f^*(I_x) \subset I_y$$

hence

$$K_x \subset \bigcap_{y \in f^{-1}(x)} K_y$$

and therefore $Y_x(\Phi) = f^{-1}(x)$ for each $x \in X$ so that by Remark 2.10 we have

$$\tilde{\Phi} : f^*(\xi) \longrightarrow {}_Y\xi_M \text{ over } Y,$$

is unique so that

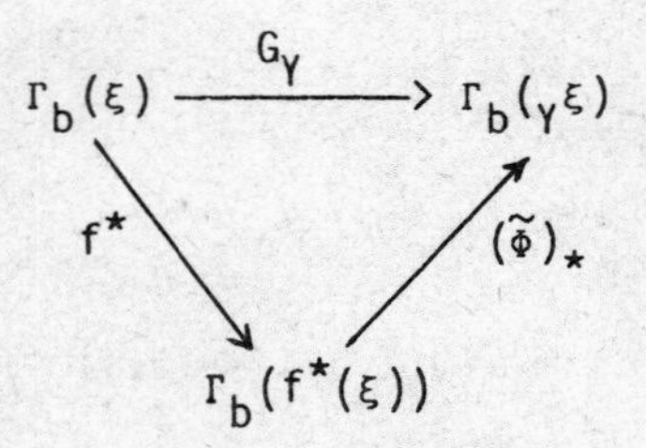

is a commutative diagram.

EXAMPLE 2.13. Suppose ξ is a C*-bundle and X is a compact Hausdorff space. Let $M = \Gamma(\xi)$ and take $\zeta = \check{\xi}_M$ and $\Phi = {}^\vee : M \to \Gamma_b(\zeta)$, as in Example 1.9. Then as $\hat{p} : \text{Prim}(M) \to X$ is continuous by Proposition 2.7, we see that $\xi_\Phi \cong \text{Prim}(\xi)$ over X and

$$\tilde{\Phi} : \hat{p}^*(\xi) \longrightarrow \check{\xi}_M$$

over Prim M is unique such that

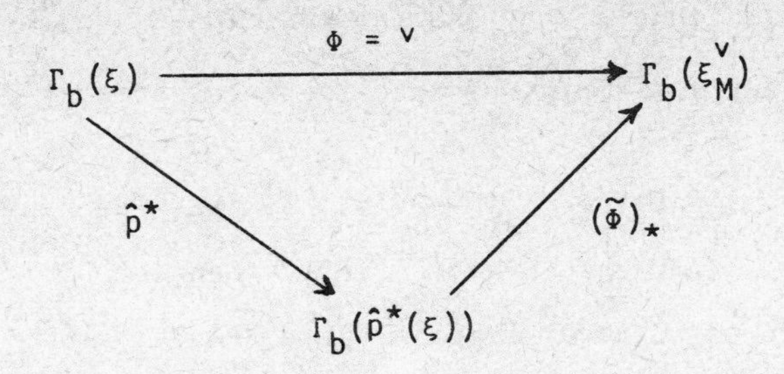

is a commutative diagram. Thus, if $P \in \text{Prim}(M)$ and $\hat{p}(P) = x$, then $K_x \subset P_\#$.

REMARK 2.14. Suppose that F is a Banach space and $\mathcal{T}$ is an auxiliary topology on F possibly different from the norm topology (for instance some weak topology). Let V be the space obtained by giving F the topology $\mathcal{T}$. If $\zeta = \varepsilon(Y,V)$ is the product bundle, then identifying each fibre of ζ with V we see that Φ just amounts to a bounded family $(ev_y \circ \Phi)$ of bounded linear maps $M \to F$ indexed by Y such that for each $m \in M$ the map $\Phi(m) : Y \to V$ is continuous. Thus $h^*(\zeta) \cong \varepsilon(E(\Phi); V)$ over $E(\Phi)$ and we can consider $\tilde{\Phi} : p_\Phi^*(\xi) \to \varepsilon(E(\Phi); V)$. By 2.11 we know $Y(\Phi)$ is closed in Y if X is compact. In particular, we can take Y to be a bounded subset of $L(M,F)$ and give Y the subspace topology from the inclusion $Y \subset V^M$. We can then define $\Phi : M \to \Gamma_b(\zeta)$ by

$$ev_y \circ \Phi = y : M \longrightarrow F.$$

Set $Y_X = Y(\Phi)$ here. Then Y_X is the subset of Y consisting of maps which factor through evaluations of ξ and this is closed in Y by Remark 2.11 if X is compact. Several examples of this are interesting.

EXAMPLE 2.15. If ξ is a C*-bundle, X is compact, and M is a C*-bundle, and M is a C*-subalgebra of $\Gamma(\xi)$, then let $Y = S(M) \subset M^d$ be the state space of M, so we take $F = V = \mathbb{C}$ in Remark 2.14. Every pure state of M extends to a pure state of $\Gamma(\xi)$ and thus factors through a point evaluation by Theorem 2.4. Thus $S(M)_X$ is closed in $S(M)$ and contains the pure states. If M has identity, then $S(M)$ is itself compact, so $S(M)_X$ is compact.

EXAMPLE 2.16. If ξ is a C*-bundle, X is compact, and M is a C*-subalgebra of $\Gamma(\xi)$, and $F = V$ is a C*-algebra, then take $Y = \mathrm{Hom}(M,F)$ which is a bounded set of bounded linear maps. Thus Y_X the set of *-homomorphisms $M \to F$ which factor through point evaluations is point-norm closed in $\mathrm{Hom}(M,F)$, and $\tilde{\Phi} : p_{\Phi}^{*}(\xi) \to \varepsilon(Y,F)$ is a C*-bundle map.

EXAMPLE 2.17. This is the same as Example 2.16, except we take $F = \mathcal{B}(H)$ where H is a Hilbert space and take V to be $\mathcal{B}(H)$ in its weak operator topology. Then $Y = \mathrm{Hom}(M,F) = \mathrm{Rep}(M,H) \subset V^M$ and $Y_X = \mathrm{Rep}_X(M,H)$ is the set of representations which factor through point evaluations and is closed in $\mathrm{Rep}(M,H)$ by Remark 2.11.

In what follows, if A is a C*-algebra, then $S_0(A)$ will denote the set of positive linear functionals in the unit ball of the dual of A, we let $S(A)$ denote the set of states of A and $P(A)$ the set of pure states of A.

Thus $P(A) \subset S(A) \subset S_0(A)$. We equip $S_0(A)$ with the weak*-topology. $S_0(A)$ is compact and $\{0\} \cup P(A)$ is the set of extreme points of $S_0(A)$ by 2.5.5 of [4].

PROPOSITION 2.17. *Let ξ be a C*-bundle over the compact Hausdorff space X. Let A be a C*-subalgebra of $\Gamma(\xi)$ which is full for ξ and let $S_0(A)_X$ denote the set of members of $S_0(A)$ which factor through point evaluations. Then $S_0(A)_X$ is compact, contains $\{0\} \cup P(A)$, and $S_0(A)$ is the closed convex hull of $S_0(A)_X$.*

Proof. By Remark 2.11 we know that $S_0(A)_X$ is closed in $S_0(A)$ and as $S_0(A)$ is compact, $S_0(A)_X$ is also. Obviously $0 \in S_0(A)_X$. If $f \in P(A)$, then f extends to a pure state of $\Gamma(\xi)$ by 2.10.2 of [4] and hence factors through a point evaluation by Theorem 2.4 and the uniqueness of Theorem 2.6 allowing us to identify $\xi_B = \xi$ where $B = \Gamma(\xi)$. Thus $\{0\} \cup P(A) \subset S_0(A)_X$. Let K be the closed convex hull of $S_0(A)_X$. Then K is a compact subset of of $S_0(A)$ because $S_0(A)_X$ is compact and hence by the Krein-Milman theorem and 2.5.5. of [4] we know $S_0(A) = K$. □

The following proposition is a useful Stone-Weierstrass theorem for C*-algebras of sections.

PROPOSITION 2.18. *Let ξ be a C*-bundle over the compact Hausdorff space X and suppose we are given C*-subalgebras $B \subset A \subset \Gamma(\xi)$. Suppose either (1) or (2) below holds:*

(1) $A|Y = B|Y$ *whenever* $Y \subset X$ *such that* Y *has cardinality* ≤ 2,

(2) B *separates* $\{0\} \cup P(A)$ *and for each* $x \in X$ *we have* $B(x) = A(x)$. *Then* $A = B$.

Proof. Let ζ be the C*-subbundle of ξ such that $\zeta(x) = A(x)$ for each $x \in X$, given to us by Proposition 1.3. After replacing ξ by ζ if necessary, we can assume that A is full for ξ. Hypothesis (1) guarantees that B separates the points of $S_0(A)_X$. In paticular, (1) implies (2). Since $S_0(A)_X$ of Proposition 2.17 contains $\{0\} \cup P(A)$ and is compact it must also contain $\{0\} \cup \overline{P(A)}$. At this point we apply the Stone-Weierstrass-Glimm theorem, 11.5.2 of [4] and conclude that if (1) holds then $A = B$, just as Fell did in Theorem 1.4 of [18]. However, using our developments so far and the following lemma we can give an elementary proof which shows that (2) implies (1), completing the proof of the proposition.

LEMMA 2.19. *Suppose* $h : A \to C_1 \times \ldots \times C_n$ *is a* *-*homomorphism of* A *into a product of* C*-*algebras where* $h = (h_1, \ldots, h_n)$, $h_i : A \to C_i$, $1 \leq i \leq n$. *Suppose that* B *is a* C*-*subalgebra of* A *which separates* $\{0\} \cup P(A)$ *and for which* $h_i(B) = h_i(A)$, $1 \leq i \leq n$. *Then* $h(A) = h(B)$.

Before giving the proof of Lemma 2.19, we use it to complete the proof of Proposition 2.18. So we are back inside $\Gamma(\xi)$. Let $Y = \{x_1, \ldots, x_n\}$ be an arbitrary finite subset of X and apply Lemma 2.19 with h_i being the evaluation at x_i for $1 \leq i \leq n$. Thus we conclude $A|Y = B|Y$ if (2) holds and in particular, (2) implies (1).

Our proof of Lemma 2.19 will be based on the following lemma.

LEMMA 2.20. *Let* A *be a* C*-*algebra and* B *a* C*-*subalgebra of* A *which separates* $\{0\} \cup P(A)$. *Let* I *and* J *be ideals of* A. *Then* $(I + J) \cap B = (I \cap B) + (J \cap B)$.

Proof. Even if A already has an identity, we choose an identity $1 \notin A$ and adjoin it to A obtaining the C*-algebra $A^{(1)}$. Let $B^{(1)} = B + \mathbb{C}1 \subset A^{(1)}$. Then A is an ideal of $A^{(1)}$ and $B^{(1)} \cap A = B$. By 2.11.8 of [4] the fact that B separates $\{0\} \cup P(A)$ will guarantee that $B^{(1)}$ separates $P(A^{(1)})$. Thus by 11.4.1 of [4] we can conclude

$$(I + J) \cap B^{(1)} = (I \cap B^{(1)}) + (J \cap B^{(1)}).$$

But since $B^{(1)} \cap A = B$ it follows that $B^{(1)} \cap K = B \cap K$ for any ideal K of B. □

We now mimic the proof of 11.4.2 of [4] using Lemma 2.20 in place of 11.4.1 of [4] and obtain Lemma 2.19 in the case n = 2. In the general case, let $K_i = \mathrm{Ker}_i h$ and observe that the conclusion of Lemma 2.19 is equivalent to

$$A = B + (K_1 \cap K_2 \cap \dots \cap K_n)$$

Working inductively, we can assume $A = B + K_0$ where $K_0 = K_1 \cap \dots \cap K_{n-1}$, but then as $A = B + K_n$, we can conclude $A = B + K_0 \cap K_n$ by applying the case where $n = 2$. □

In case A is a C*-algebra such that Prim A is Hausdorff, then by 3.3.9 of [4] and Proposition 1.3 we know (see Example 1.9) that $P_\# = P$ for each $P \in \mathrm{Prim}\, A$ so that $\xi_A^\vee = \xi_A^{01} = \xi_A^0$. Thus $\wedge : A \to \Gamma_0(\xi_A^\vee)$ is a *-homomorphism,

$\wedge(A)$ is full for $\xi_A^\vee$, and $\xi_A^\vee(P) = A/P$ for $P \in \text{Prim } A$. Let ξ be the C*-bundle over X, the one-point compactification of Prim A, whose fibre over the point at infinity is just zero and with $\xi|\text{Prim } A = \xi_A^\vee$. Then $\Gamma_0(\xi_A^\vee) \cong \Gamma(\xi)$ and we can identify $\wedge$ with a *-homomorphism $\wedge : A \to \Gamma(\xi)$ which is injective. By the Dauns-Hofmann theorem $\wedge(A)$ is a $C_b(\text{Prim } A)$-submodule of $\Gamma_0(\xi_A^\vee)$ and is therefore a $C(X)$-submodule of $\Gamma(\xi)$ which is full for ξ.
But then by Proposition 2.3 we have $\wedge(A) = \Gamma(\xi)$. Thus $\wedge : A \cong \Gamma_0(\xi_A^\vee)$ when Prim A is Hausdorff and we can identify $\xi_A^\vee = {}_X\xi_A$ by the uniqueness of Theorem 2.6. This is Fell's construction [18]. In particular, this means that Theorem 2.4 can be viewed as a generalization of Theorem 2.3 in [18] to the case where Prim B is possibly non-Hausdorff.

As already remarked, we can use the functors R_X and Γ_X to make module valued functors into Banach bundle valued functors. If M is a functor of n variables with values in $\mathcal{M}(Y)$ and the i-th variable in $\mathcal{M}(Y_i)$, then set

$$R(M) = R_Y \circ M \circ (\Gamma_{Y_1} \times \dots \times \Gamma_{Y_n}).$$

Then for the functor $\text{Hom}_{C_b(Y)}(\_,\_)$ the result is (isometrically) isomorphic over Y to $\text{HOM}_Y(\_,\_)$ via the isomorphism (2.2), so the result is not generally useful, as we noted earlier, unless the first variable is locally trivial with finite dimensional fibre. This is a useful construction, when dealing with multiplier algebras of section algebras. In particular, if ξ is a C^*-bundle over the locally compact Hausdorff space Y and if

$A = \Gamma_0(\xi)$ is the C^*-algebra of sections vanishing at infinity, then as mentioned at the end of Chapter 1, we can get a unique C^*-bundle ζ over X where X is the Stone-Cech compactification of Y, with $\zeta|Y = \xi$ and $\zeta|X\setminus Y = 0$. Then $\Gamma(\zeta) \cong \Gamma_0(\xi)$ under the restriction map, hence $M(A) \cong \Gamma(M(\zeta))$. Thus, we need to include points at infinity in constructing the C^*-bundle which corresponds to $M(A)$.

We can also use this technique of transferring module functors to bundles to define push forwards for Banach and C^*-bundles. If $f:X \to Y$ is a continuous map, then $f^* : C_b(Y) \to C_b(X)$ is a unital $*$-homomorphism so if M is a Banach $C_b(X)$-module (respectively, a central $C_b(X)$-$*$algebra), then let f_*M denote the associated Banach $C_b(Y)$-module (respectively, central $C_b(Y)$-$*$algebra) obtained by the $*$-homomorphism $f^* : C_b(Y) \to C_b(X)$. Thus, ξ_{f_*M} is a Banach (respectively C^*-) bundle over Y, and in this case we put $G(f_*M) = f_*$ so $f_* : f_*M \to \Gamma_b(\xi_{f_*M})$. If ξ is a Banach (respectively, C^*-) bundle over X, then we define

$$f_*\xi = \xi_{f_*\Gamma_b(\xi)}$$

so that $f_*\xi$ is a Banach (respectively, C^*-) bundle over Y, and

$$f_* : f_*\Gamma_b(\xi) \to \Gamma_b(f_*\xi)$$

is a contractive $C_b(Y)$-homomorphism which is necessarily an isometric $C_b(Y)$-isomorphism, if Y is compact, because $f_*\Gamma_b(\xi)$ is $C_b(Y)$-convex since $\Gamma_b(\xi)$ is $C_b(X)$-convex. Thus, $f_* : \Gamma_b(\xi) \to \Gamma_b(f_*\xi)$ is linear (respectively, a $*$-homomorphism) and

$$f_*((g \circ f)s) = gf_*(s), \; g \in C_b(Y), \; s \in \Gamma_b(\xi). \qquad (2.11)$$

On the other hand, for any Banach (respectively, C^*-) bundle over Y, say ζ,

$$f^* : \Gamma_b(\zeta) \to \Gamma_b(f^*(\zeta))$$

is linear (respectively, a *-homomorphism), and satisfies

$$f^*(gt) = (g \circ f)f^*(t), \; g \in C_b(Y), \; t \in \Gamma_b(\zeta), \qquad (2.12)$$

so combining (2.11) with (2.12) for $\zeta = f_*\xi$, we see that

$$f^* \circ f_* : \Gamma_b(\xi) \to \Gamma_b(f^*(f_*\xi)) \qquad (2.13)$$

is $f^*(C_b(Y))$-linear. Suppose X and Y are compact Hausdorff spaces, and set $X_0 = f(X) \subset Y$. If $y \in Y \backslash X_0$, then by complete regularity of Y, we have $f^*(I_y) = f^*(C_b(Y))$. Suppose in addition that $f^* : C(Y) \to C(X)$ is surjective. We now have for $y \in Y \backslash X_0$, $I_y[f_*\Gamma(\xi)] = C(X)\Gamma(\xi) = \Gamma(\xi)$, and hence

$$(f_*\xi)(y) = 0, \; y \in Y \backslash X_0, \qquad (2.14)$$

and thus

$$\|s|X_0\| = \|s\|, \;\; s \in \Gamma(f_*\xi). \qquad (2.15)$$

On the other hand, as $f(X) = X_0$, we have $\|f^*(s)\| = \|s|X_0\|$ for $s \in \Gamma(f_*\xi)$, and therefore $f^* : \Gamma(f_*\xi) \to \Gamma(f^*(f_*\xi))$ is isometric. But $f_* : f_*\Gamma(\xi) \to \Gamma(f_*\xi)$ is an isometric $C(Y)$-isomorphism, as Y is compact and $f_*\Gamma(\xi)$ is $C(Y)$-convex; thus $f^* \circ f_*$ of (2.13) is now an isometric $C(X)$-isomorphism, so by Theorem 2.6,

$$f^{*} \circ f_{*}(\xi) \cong \xi \tag{2.16}$$

over X, when f^{*} is surjective and X and Y are compact Hausdorff. Of course, now f is actually a homeomorphism onto $X_0 \subset Y$. In particular, this shows that if $f = i : X \subset Y$ is an inclusion map with Y compact Hausdorff and X closed in Y, then combining (2.14) and (2.16), we have

$$\begin{aligned} i_{*}(\xi)|X &\cong \xi \quad \text{over } X \\ [i_{*}(\xi)](y) &= 0, \qquad y \in Y \backslash X. \end{aligned} \tag{2.17}$$

This example is of some interest, as it shows that Theorem 3.5 of [13] cannot hold for general C*-bundles (in [13] the term C*-bundle means (F)C*-bundle). In fact, no (F)Banach bundle ξ over Y can satisfy (2.17) if ξ has all non-zero fibres [10, Proposition 1.6], unless X is open in Y, which is usually not true as X is already closed. Thus, assume that X is closed but not open in Y. In particular, if $\xi = \varepsilon(X,F)$ with $F \neq 0$, then ξ is an (F)Banach bundle over X, but $i_{*}(\xi)$ is not an (F)Banach bundle over Y, although it is over X, by (2.17). Even though it may seem that this example is somehow trivial, it shows that the local structure of general Banach bundles and C*-bundles behaves rather pathologically in comparison to that for (F)Banach bundles and (F)C*-bundles. For instance, if we have a direct sum $\varepsilon_0 = [i_{*}(X,F_1)] \oplus \varepsilon(Y,F_2)$, say with F_1 and F_2 both Hilbert spaces, $F_1 \neq 0 \neq F_2$, so that $\varepsilon_0(x) \cong F_1 \oplus F_2$, when $x \in X$, $\varepsilon_0(x) = F_2$ when $x \in Y \backslash X$, then ε_0 is a Hilbert bundle but not an (F)Hilbert bundle, and Hilbert bundles locally isomorphic to ε_0 cannot be classified by any of the methods developed in [10] or [11] for (F)Hilbert bundles.

Our final remarks about Theorem 2.6 will serve to put it in perspective with classical complex vector bundle theory. The reader may have noticed our development here in many ways parallels that used in [1, Appendix] to prove Swan's Theorem [1, page 735]. In fact, Theorem 2.6 is a generalization of the preliminary result [1, (2.4), page 734], and Swan's theorem can be quickly obtained from our Theorem 2.6 in the following parallel way. Let X be a compact Hausdorff space and $Z = C(X)$. For n a positive integer, $Z^n \cong \Gamma(\varepsilon(X:\mathbb{C}^n))$ so Z^n has a natural Hilbert Z-module structure. If $\phi : Z^n \to Z$ is a Z-homomorphism, and if $e_i = (\delta_{ij}) \in Z^n$, $1 \leq i \leq n$, then set $s_i = \phi(e_i)$ and define $\psi : \varepsilon(X; \mathbb{C}^n) \to \varepsilon(X; \mathbb{C})$ by

$$\psi(x,(z_1,\dots,z_n)) = (x, \sum_1^n z_i s_i(x)).$$

Then $\psi_* = \phi$ under the identification $Z^n = \Gamma(\varepsilon(X; \mathbb{C}^n))$. It follows that Z-module homomorphisms of finitely generated (henceforth, f.g.) free Z-modules are necessarily bounded. If M is an f.g. projective Z-module, then M is a direct summand of an f.g. free Z-module, say Z^n. Thus, $M \cong \operatorname{Ker} \phi$ for some idempotent Z-homomorphism $\phi : Z^n \to Z^n$, and as ϕ is bounded, it follows that M is a Hilbert Z-submodule of Z^n, and that $\xi_M \subset \varepsilon(X; \mathbb{C}^n)$ with $\operatorname{Ker} \phi = \Gamma(\xi_M)$. Likewise, with $\eta \subset \varepsilon(X; \mathbb{C}^n)$ so that $\operatorname{Ker}(1 - \phi) = \Gamma(\eta)$, we can apply [10, Proposition 1.6] to both ξ_M and η to conclude that ξ_M and η are homogeneous of finite rank and hence locally trivial [10, Prop. 2.3]. Thus, under Theorem 2.6, the f.g. projective Z-modules arise from locally trivial Hilbert bundles having finite dimensional fibre, as every such Hilbert bundle is a direct summand of a trivial bundle with finite dimensional fibre [29].

In particular, the isomorphism classes of Hilbert line bundles over X under the tensor product $\otimes_X$ form a group $\mathrm{Pic}(X)$, the classical Picard group. And by Swan's theorem, $\mathrm{Pic}(X)$ is isomorphic to $\mathrm{Pic}\ Z$, the group of invertible f.g.p. Z-modules under $\otimes_Z$. Now, let $\mathcal{S}_Y$ denote the sheaf of germs of Y-valued continuous maps. Then

$$0 \longrightarrow \mathcal{S}_{\mathbb{Z}} \longrightarrow \mathcal{S}_{\mathbb{R}} \xrightarrow{\exp 2\pi i(-)} \mathcal{S}_{S^1} \longrightarrow 1$$

is a (locally) exact sequence of sheaves whose long exact cohomology sequence gives a coboundary isomorphism (all $n > 0$)

$$\delta_n : H^n(X;\mathcal{S}_{S^1}) \cong H^{n+1}(X;\mathcal{S}\ ) \cong \check{H}^{n+1}(X;\mathcal{S})$$

because $\mathcal{S}$ is a fine sheaf. Here $\check{H}$ is the Čech cohomology functor. In particular, taking $n = 1$, gives the isomorphism
$\mathrm{Pic}(X) \cong \check{H}^1(X;\mathcal{S}_{S^1}) \cong \check{H}^2(X;\mathbb{Z})$ which maps a Hilbert line bundle to its first Chern class. These isomorphisms will be used repeatedly in what follows, as we generalize to Picard groups of non-commutative C^*-algebras.

3 Automorphisms of a C*-algebra and Hilbert submodules

Let A be a C^*-algebra with identity 1 and center Z. Let $U(A)$ denote the unitary group of A, let $\mathrm{Aut}\, A$ denote the automorphism group of A, and let $\mathrm{Aut}_Z A$ denote the group of Z-automorphisms of A; that is, the subgroup of $\mathrm{Aut}\, A$ consisting of those automorphisms which are also Z-module automorphisms when A is given its natural Z-module structure. Thus $\alpha \in \mathrm{Aut}\, A$ belongs to $\mathrm{Aut}_Z A$ if and only if $\alpha(z) = z$ for each $z \in Z$. Let $\mathrm{Inn}\, A$ be the inner automorphism group of A, and for $u \in U(A)$, let $\mathrm{Ad}(u)$ denote the inner automorphism $a \longrightarrow uau^*$. There is an exact sequence of topological groups

$$1 \longrightarrow U(Z) \longrightarrow U(A) \xrightarrow{\mathrm{Ad}} \mathrm{Inn}\, A \longrightarrow 1$$

where each group carries the norm topology. Obviously $\mathrm{Inn}\, A$ is a subgroup of $\mathrm{Aut}_Z A$, and it follows from the identity $\alpha \circ \mathrm{Ad}(u) = \mathrm{Ad}(\alpha(u)) \circ \alpha$, for $u \in U(A)$ and $\alpha \in \mathrm{Aut}\, A$, that $\mathrm{Inn}\, A$ is a normal subgroup of $\mathrm{Aut}_Z A$. So we can inquire as to the structure of the quotient group $\mathrm{Aut}_Z A/\mathrm{Inn}\, A$.

For each $\alpha \in \mathrm{Aut}_Z A$, let M_α be the Z-submodule of A defined by

$$M_\alpha = \{m \in A : \alpha(a)m = ma \text{ for each } a \in A\} .$$

It is routine to verify that

$$M_\alpha M_\beta \subset M_{\alpha\beta} , \tag{3.1}$$

$$M^*_\alpha = M_{\alpha^{-1}} , \tag{3.2}$$

and therefore, in particular,

$$M_\alpha^* M_\alpha \subset Z \supset M_\alpha M_\alpha^* .$$

This last observation means that M has a natural Hilbert Z-module structure given by

$$\langle m|n\rangle = n^* m \in Z \qquad (m,n \in M_\alpha) .$$

Let us agree to call M a *Hilbert Z-submodule* of A if M is a closed linear subspace of A which is also a Z-submodule of A satisfying $M^*M \subset Z \supset MM^*$. Notice that M^* is then also a Hilbert Z-submodule of A. From the previous chapter, we know that there is a uniquely determined Hilbert Bundle ξ_M over $X = \hat{Z}$ such that M and $\Gamma(\xi_M)$ are isometrically isomorphic Hilbert Z-modules. (*In this section all Hilbert bundles are understood to be* (F) *Hilbert bundles unless explicitly stated otherwise*). Thus each $\alpha \in \mathrm{Aut}_Z A$ determines a Hilbert Z-submodule M of A and a Hilbert bundle ξ_α (where $\xi_\alpha = \xi_M$ with $M = M_\alpha$).
We let $[\xi]$ denote the isomorphism class (over X) of the Banach bundle ξ and we let $[M]$ denote the isometric Z-isomorphism class of the Banach Z-module M. In view of the results of Chapter 2, we can without any real confusion write $[\xi] = [\Gamma(\xi)]$. We write $[\xi] < [\zeta]$ if $\xi < \zeta$.
If M and N are Hilbert Z-submodules of A, let $M \circ N$ be the closed linear span of $MN = \{mn : m \in M , n \in N\}$.

LEMMA 3.1. *Let* M *and* N *be Hilbert Z-submodules of* A. *Then* $M \circ N$ *is also a Hilbert Z-submodule of* A.

Proof. First, $M \circ N$ is a Z-submodule of A because both ZMN and MNZ are contained in MN. Using $M^*M \subset Z$ and $N^*N \subset Z$ we have

$$(MN)^*(MN) = N^*M^*MN \subset N^*ZN = ZN^*N \subset Z .$$

Similarly, $(MN)(MN)^* \subset Z$. It follows that

$$(M\circ N)^*(M\circ N) \subset Z \subset (M\circ N)(M\circ N)^* . \qquad \square$$

If ξ and ζ are Hilbert bundles over X, then $\xi \otimes_X \zeta$ is again a Hilbert bundle over X, and up to isometric Z-isomorphism, we can take $\Gamma(\xi) \otimes_Z \Gamma(\zeta) = \Gamma(\xi \otimes_X \zeta)$ as defining the tensor product over Z of Hilbert Z-modules (where $X = \hat{Z}$). We recall from [10, Sect. 4] that if $x_i \in \Gamma(\xi)$, $y_i \in \Gamma(\zeta)$, $1 \leq i \leq k$, then

$$|\Sigma x_i \otimes y_i|^2 = \langle \Sigma x_i \otimes y_i \mid \Sigma x_i \otimes y_i \rangle = \sum_{ij} \langle x_i | x_j \rangle \langle y_i | y_j \rangle , \qquad (3.3)$$

and the set of sections of the form $\Sigma x_i \otimes y_i$ is total for $\xi \otimes_X \zeta$. We let $\mathring{\otimes}_Z$ denote the algebraic tensor product operation for Z-modules. We obviously then have a canonical Z-homorphism

$$\Gamma(\xi) \mathring{\otimes}_Z \Gamma(\zeta) \xrightarrow{t} \Gamma(\xi) \otimes_Z \Gamma(\zeta)$$

with dense image. (The interested reader will note that our definition $\Gamma(\xi) \otimes_Z \Gamma(\zeta) = \Gamma(\xi \otimes_X \zeta)$ agrees with the definition of [39, Theorem 5.9] in the special case $A = B = C = Z$.)

LEMMA 3.2. *Let* M *and* N *be Hilbert* Z-*submodules of* A. *If* $m_i \in M$ *and* $n_i \in N$, $1 \leq i \leq k$, *then*

$$|\Sigma m_i \otimes n_i| = |\Sigma m_i n_i|$$

where the left-hand side is calculated in $M \otimes_Z N = \Gamma(\xi_M) \otimes_Z \Gamma(\xi_N) = \Gamma(\xi_M \otimes_X \xi_N)$ *and the right-hand side is calculated in* $M \circ N$.

Proof. By equation (3.3) and the fact that $m_j^* m_i \in Z$, for all i,j,

$$|\sum_{i=1}^{k} m_i \otimes n_i|^2 = \sum_{i,j=1}^{k} \langle m_i | m_j \rangle \langle n_i | n_j \rangle = \sum_{i,j=1}^{k} (m_j^* m_i)(n_j^* n_i)$$

$$= \sum_{i,j=1}^{k} n_j^*(m_j^* m_i) n_i = \sum_{i,j=1}^{k} (m_j n_j)^* (m_i n_i)$$

$$= \sum_{i,j=1}^{k} \langle m_i n_i | m_j n_j \rangle = \langle \sum_{i=1}^{k} m_i n_i | \sum_{j=1}^{k} m_j n_j \rangle$$

$$= |\sum_{i=1}^{k} m_i n_i|^2 . \qquad \square$$

We now notice that if M and N are Hilbert Z-submodules of A, then the map $(m,n) \to mn$ is a Z-bilinear map of $M \times N$ into $M \circ N$, so there is a Z-linear map $b : M \mathring{\otimes}_Z N \to M \circ N$ of Z-modules. If $t : M \mathring{\otimes}_Z N \to M \otimes_Z N$ is the canonical linear map, then by Lemma 3.2 we have $|b(a)|\ |t|t(a)|$, for $a \in M \otimes_Z N$. It follows that if $t(a) = 0$, then $b(a) = 0$, and hence, since t has dense range, there is a unique Z-linear map $\overline{b}: M \otimes_Z N \to M \circ N$ such that $\overline{b} \circ t = b$. Of course $\|\overline{b}(s)\| = \|s\|$ for each $s \in M \otimes_Z N$, and thus $|\overline{b}(s)| = |s|$ for each $s \in M \otimes_Z N$; that is, we have $[M \otimes_Z N] \leq [M \circ N]$. But since $M \circ N$ is the Hilbert Z-submodule of A generated by the image of the map $(m,n) \to mn$, the image of $\overline{b}$ is dense in $M \circ N$. It follows that $\overline{b}$ is an isometric Hilbert Z-module isomorphism of $M \otimes_Z N$ onto $M \circ N$. This proves our next result.

PROPOSITION 3.3 *If* M *and* N *are Hilbert* Z-*submodules of* A, *then* $[M \otimes_Z N] = [M \circ N]$. $\square$

COROLLARY 3.4. *If* M *is a Hilbert* Z-*submodule of* A, *then* ξ_M *has fibres only of dimension* 0 *or* 1; *in fact* $\xi_M \otimes_X \xi_{M^*} \leq \varepsilon^1$, *where ε^1 denotes the trivial line bundle over X. Thus* $[M \otimes_Z M^*] \leq [Z]$.

Proof. We simply apply Proposition 3.3 to M and $N = M^*$, keeping in mind that since $MM^* \subset Z$, we have $M \circ M^* \subset Z$ and hence $[M \otimes_Z M^*] \leq [M \circ M^*] \leq [Z] = [\varepsilon^1]$. □

Let $H(A)$ denote the set of isometric Z-isomorphism classes of Hilbert Z-submodules of A. We make $H(A)$ into a semigroup by defining $[M][N] = [M \circ N]$. We see by Proposition 3.3 that this is equivalent to setting $[M][N] = [M \otimes_Z N]$. Moreover, $H(A)$ has a partial order $\leq$, as already defined. If ξ is a Hilbert bundle all of whose fibres are of dimension 0 or 1, then ξ is a (0,1)-bundle [11, Sect. 5]. We let $\mathrm{supp}(\xi)$ be the set of points over which ξ has non-zero fibres. Then $\mathrm{supp}(\xi)$ is open in X (by norm continuity). In view of [11, Lemma 3.2, p. 253] and the remarks preceding Proposition 4.4 of [11, p. 265], we see that if ξ_* is the conjugate bundle [11, p. 264], then $\xi_* \otimes_X \xi$ is an elementary (0,1)-bundle [12, p. 314]. Thus $\xi_* \otimes_X \xi \leq \varepsilon^1$, where ε^1 is the trivial Hilbert line bundle over X. We can therefore form the partially ordered commutative semigroup $H_0^1(X)$ of isomorphism classes of (0,1)-bundles over X under the tensor product operation $[\xi][\zeta] = [\xi \otimes_X \zeta]$. Then $[Z] = 1$ in $H_0^1(X)$ and thus $1 \in H(A) \subset H_0^1(X)$. Moreover, the idempotents of $H_0^1(X)$ are exactly the elements $[\varepsilon]$, where ε is an elementary (0,1)-bundle over X, and the subgroup $G_{[\varepsilon]}(X)$ of elements of $H_0^1(X)$ having inverses with respect to $[\varepsilon]$ are exactly the (0,1)-bundles having the same support as ε; so $G_{[\varepsilon]}(X) = \check{H}^2(\mathrm{supp}(\varepsilon), \mathbb{Z})$, the second Čech cohomology group. Since $\xi_{M^*} = (\xi_M)_*$ is easily verified, if $[\varepsilon] \in H(A)$, then $G_{[\varepsilon]}(X) \cap H(A)$ is exactly the subgroup of $H(A)$ of elements having inverse with respect to $[\varepsilon]$. Of course by Proposition 3.3, $H(A)$ is just a subsemigroup of $H_0^1(X)$.

Now define $h: \mathrm{Aut}_Z A \to H(A)$ by the rule $h(\alpha) = [M_\alpha]$. In view of the inequality (3.1), the following super-multiplicative property holds

$$h(\alpha)h(\beta) \leq h(\alpha\beta) \qquad (\alpha,\beta \in \mathrm{Aut}_Z A) . \tag{3.4}$$

Suppose that $[M] = [Z]$ with M a Hilbert Z-submodule of A. Then by Theorem 2.6, there is an isometric Z-isomorphism $T: Z \cong M$ such that $|T(z)| = |z|$, for each $z \in Z$. In particular, setting $T(1) = u$, we have $|u| = 1$. Hence in A ,

$$u^*u = \langle u|u\rangle = |u|^2 = 1 ,$$

so u is partial isometry and for each $z \in Z$, $T(z) = zT(1) = zu$. Thus $Zu = M$. On the other hand, as $MM^* \subset Z$ also, we have $uu^* \in Z$ and thus

$$uu^* = (uu^*)(u^*u) = u^*(uu^*)u = 1^2 = 1 ,$$

and therefore $u \in U(A)$. This proves the next result.

PROPOSITION 3.5. _IF_ M _is a Hilbert Z-submodule of_ A _with_ $[M] = [Z]$ _in_ $H(A)$, _then there is a unitary_ $u \in M \cap U(A)$ _with_ $M = uZ$. □

THEOREM 3.6. _The following sequence is exact:_

$$1 \longrightarrow \mathrm{Inn}\ A \longrightarrow \mathrm{Aut}_Z A \xrightarrow{h} H(A) .$$

Proof. Suppose $u \in U(A)$ and put $\alpha = \mathrm{Ad}(u)$. Then for any $a \in A$, $\alpha(a)u = uau^*u = ua$, which means that $u \in M_\alpha$, and therefore $u^*M_\alpha = Z$.

But then $M_\alpha = uZ$ and the map $\phi: Z \to M_\alpha$ given by $\phi(z) = zu$ is obviously an isometric Z-isomorphism, and therefore $[M_\alpha] = [Z]$ which means $h(\alpha) = 1$ in $\mathcal{H}(A)$. Conversely, suppose that $\alpha \in \mathrm{Aut}_Z A$ and $h(\alpha) = 1$ in $\mathcal{H}(A)$. By Proposition 3.5 there is a unitary $u \in M_\alpha \cap U(A)$ with $uZ = M_\alpha$. But then $u \in M_\alpha$ so for every $a \in A$, we have $\alpha(a)u = ua$, and therefore $\alpha = \mathrm{Ad}(u)$. □

DEFINITION 3.7. <u>By</u> Pic(A,Z) <u>we mean the group of units of</u> $\mathcal{H}(A)$. <u>Define the subgroup</u> KInn A <u>of</u> $\mathrm{Aut}_Z A$ <u>by</u>

$$\mathrm{KInn}\ A = h^{-1}(\mathrm{Pic}(A,Z))$$

<u>and let</u> h_1 <u>be the restriction of</u> h <u>to</u> KInn A.

It is clear fromt the discussion following Corollary 3.4 that Pic(A,Z) is the group of equivalence classes $[M] \in \mathcal{H}(A)$ of Hilbert Z-submodules M with $\mathrm{supp}\ \xi_M = X$. With these definitions, the following lemma is an immediate consequence of Theorem 3.6.

LEMMA 3.8. <u>The sequence</u>

$$1 \longrightarrow \mathrm{Inn}\ A \longrightarrow \mathrm{KInn}\ A \xrightarrow{h_1} \mathrm{Pic}(A,Z)$$

<u>is an exact sequence of groups and homomorphisms</u>. □

We shall show h_1 is surjective, but first we need to look more closely at KInn A using the representation $A \cong \Gamma(\xi_A)$ given by Theorem 2.4 in which ξ_A is a C^*-bundle over $X = \hat{Z}$. For convenience we simply identify $A = \Gamma(\xi_A)$. According to the remark following Theorem 2.6, if M is a Hilbert Z-submodule of A, then $\xi_M \cong \xi'_M \subset \xi_A$, where ξ'_M is the unique Banach subbundle of ξ_A given by

$$\xi_M'(x) = M(x) = \{m(x) : m \in M\} .$$

Henceforth we just identify $\xi_M \subset \xi_A$. Notice that the condition $M^*M \subset Z \supset MM^*$ when evaluated at x in X is just

$$\xi_M(x)^*\xi_M(x) \subset \mathbb{C}\, 1(x) \supset \xi_M(x)\xi_M(x)^*.$$

If ξ is a Banach subbundle of ξ_A satisfying

$$\xi(x)^*\xi(x) \subset \mathbb{C}\, 1(x) \supset \xi(x)\xi(x)^* , \quad (x \in X) \tag{3.5}$$

then for v in $\xi(x)$ we have $(v^*v)^{1/2} = \|v\|\, 1(x)$, and we see that ξ has a unique Hilbert space structure on each fibre so that if $u,v \in \xi(x)$, then $\langle u|v\rangle 1(x) = v^*u \in \mathbb{C}\cdot 1(x)$. Moreover, if $u \in \Gamma(\xi)$, then $|u| = (u^*u)^{1/2} \in C(X)$ under the natural indentification $C(X) = C(X)1 = Z$, and hence $|u|$ is continuous, so that ξ is a Hilbert bundle. Let us agree to call ξ a <u>natural Hilbert subbundle of</u> ξ_A provided it satisfies (3.5). Thus for each Hilbert Z-submodule M of A we have a natural Hilbert subbundle ξ_M of ξ_A which is a line subbundle provided $[M]$ is in $\mathrm{Pic}(A,Z)$.

Now, by the argument used to prove Proposition 3.5, the next result is easily obtained.

LEMMA 3.9. *If ξ is a natural Hilbert subbundle of ξ_A and if $u \in \xi(x)$ is a unit vector in the fibre of ξ over $x \in X$, then u is a unitary in $\xi_A(x)$.* □

Suppose now that $\alpha \in \mathrm{KInn}\, A$. Then M_α is a Hilbert Z-submodule of A and ξ_α is an unit in $H(A)$ (where $\xi_\alpha = \xi_M$ with $M = M_\alpha$). Thus ξ_α is a natural Hilbert line subbundle of ξ_A. Also, by Theorem 2.6, as $\alpha \in \mathrm{Aut}_Z A$, we can identify α with a C^*-bundle automorphism

$$\tilde{\alpha} : \xi_A \underset{X}{\cong} \xi_A .$$

Let α_x be the automorphism of the fibre $\xi_A(x)$ determined by $\tilde{\alpha}$. As $\xi_\alpha(x)$ is a complex line in $\xi_A(x)$, we can choose a unit vector $u_x \in \xi_\alpha(x)$. By Lemma 3.9, u_x is a unitary in $\xi_A(x)$. Choose $s_x \in \Gamma(\xi_\alpha)$ with $s_x(x) = u_x$. We then have by definition of $M_\alpha = \Gamma(\xi_\alpha)$ that for every $a \in A = \Gamma(\xi_A)$, $\alpha(a)s_x = s_x a$, so evaluating at x gives $\alpha_x(b)u_x = u_x b$ for each $b \in \xi_A(x)$ because $ev_x: \Gamma(\xi_A) \longrightarrow \xi_A(x)$ is surjective. But this shows that $\alpha_x = Ad(u_x)$ for each x in X. Notice that the only requirement on u_x is that it be a unit vector in $\xi_\alpha(x)$. In particular, if y is fixed in X, choosing an open neighborhood W in y so that $s_y(x) \neq 0$ for each $x \in W$, we can choose $u_x = \|s_y(x)\|^{-1} s(x)$ for each $x \in W$ and thus locally the unitaries can be chosen continuously. If W is a closed neighborhood, then $\tilde{\alpha}|W: \xi_A|W \cong_W \xi_A|W$ and we can set

$$\alpha|W = (\tilde{\alpha}|W)_*: \Gamma(\xi_A|W) \cong \Gamma(\xi_A|W) .$$

If $u \in \Pi\xi_A$ is a unitary selection with $\alpha_x = Ad(u_x)$ for each x, we write $\alpha = Ad(u)$. We have thus shown that if $\alpha \in$ KInn A, then any x has a closed neighborhood W and a unitary section $u \in \Gamma(\xi_A|W)$ such that $\alpha|W = Ad(u)$. We denote the set of all automorphisms in $Aut_Z A$ satisfying this latter condition by LInn A and we call these automorphisms _locally inner automorphisms_. Obviously LInn A is a subgroup of $Aut_Z A$, as in PInn A, the group of all _pointwise inner automorphisms of_ A, by which we mean those $\alpha \in Aut_Z A$ for which there is a unitary in $\Pi_b \xi_A$ with $\alpha = \mathrm{Ad}(u)$. Thus

$$\mathrm{Inn}(A) \subset \mathrm{KInn}\ A \subset \mathrm{LInn}\ A \subset \mathrm{PInn}\ A \subset \mathrm{Aut}_Z(A).$$

Suppose now that $[M]$ is in $\mathrm{Pic}(A,Z)$. Thus $\xi_M \subset \xi_A$ is a natural Hilbert line subbundle. For each $x \in X$ choose a unit vector $u(x) \in \xi_M(x)$. Then $u \in \Pi_b \xi_M$ is a unitary selection of ξ_A by Lemma 3.9. Let $\alpha_x = \mathrm{Ad}(u(x)): \xi_A(x) \cong \xi_A(x)$, for all $x \in X$. Define $\alpha: \Gamma(\xi_A) \to \Pi_b \xi_A$ by setting $[\alpha(s)](x) = \alpha_x(s(x))$, for all $x \in X$, $s \in \Gamma(\xi_A)$. We claim $\alpha \in \mathrm{Aut}_Z A$; and in fact $\alpha \in \mathrm{KInn}\ A$, with $[M_\alpha] = [M]$. First notice that if $v \in \Pi_b \xi_M$ is another unitary selection of ξ_A, then as ξ_M is a complex line bundle, $\alpha_x = \mathrm{Ad}(v(x))$ for each x in X, so α is independent of the particular choice of u. To see that $\alpha \in \mathrm{Aut}_Z A$, using Theorem 2.6, we need only show that for $s \in \Gamma(\xi_A)$ we have $\alpha(s) \in \Gamma(\xi_A)$; in other words, we must show $\alpha(s)$ is continuous. If x_0 is a point X, we can find an open set W containing x_0 and a section $v \in \Gamma(\xi_M|W)$ with $|v| = 1$ on W, so v is then a unitary section of $\xi_A|W$. But then $\alpha(s)|W$ is continuous, because

$$\alpha(s)|W = v(s|W)v^*,$$

in view of our previous remark concerning the fact that α is independent of the choice of unit selection of ξ_M. So $\alpha(s)$ is locally continuous, and hence continuous. Now to show that $[M_\alpha] = [M]$, or equivalently that $h_1(\alpha) = [M]$, it suffices to show that $\Gamma(\xi_M) \subset M_\alpha$, because then $\xi_M \subset \xi_\alpha$, and since both ξ_M and ξ_α are natural Hilbert subbundles of ξ_A having support equal to X, then $\xi_M = \xi_\alpha$; i.e., $h_1(\alpha) = [M]$. So suppose that $s \in \Gamma(\xi_M)$. For each $x \in X$ there is an $f(x) \in \mathbb{C}$ with $s(x) = f(x)u(x)$. For $a \in A$ we have

$$[\alpha(a)s](x) = \alpha_x(a(x))f(x)u(x) = f(x)\alpha_x(a(x))u(x)$$
$$= f(x)u(x)a(x)u(x)^*u(x) = f(x)u(x)a(x)$$
$$= s(x)a(x)$$

for each $x \in X$, and thus $\alpha(a)s = sa$ for every $a \in A$. This shows that $\Gamma(\xi_M) \subset M_\alpha$, and completes the proof of our next lemma:

LEMMA 3.10. *The sequence*

$$1 \longrightarrow \text{Inn } A \longrightarrow \text{KInn } A \xrightarrow{h_1} \text{Pic}(A,Z) \longrightarrow 0$$

is an exact sequence of groups and homomorphisms. □

As demonstrated in [36], if ξ_A is locally trivial, the natural group to consider is LInn A. But even without assuming local triviality we can show that LInn A equals KInn A.

PROPOSITION 3.11. KInn A = LInn A.

Proof. Suppose $\alpha \in$ LInn A. To show that $\alpha \in$ KInn A we need only show that $\text{supp } \xi_\alpha = X$, or equivalently, with $x_0 \in X$ we find $s \in M_\alpha$ with $s(x_0) \neq 0$. Choose W open in X with $x_0 \in W$ and $u \in \Gamma(\xi_A|W)$ a unitary section having $\text{Ad } u = \alpha|W$. Let N be a closed neighborhood of x_0 contained in W and $f : X \to [0,1]$ a continuous map vanishing off N such that $f(x_0) = 1$. We can now define $s \in \Gamma(\xi_A)$ by $s|X\backslash N = 0$ and $s|W = fu$. Then for any $a \in A = \Gamma(\xi_A)$, we have

$$[\alpha(a)s](x) = u(x)a(x)u(x)^*f(x)u(x)$$
$$= f(x)u(x)a(x)$$
$$= s(x)a(x) = (sa)(x), \ x \in W$$

whereas, $[\alpha(a)s](x) = 0 = (sa)(x)$, $x \in X\backslash N$. Thus $\alpha(a)s = sa$ for every $a \in A$ and this by definition means $s \in M_\alpha$. But of course, $s(x_0) = u(x_0) \neq 0$. □

Combining Lemma 3.10 and Proposition 3.11 now gives our main theorem.

THEOREM 3.12. *The sequence*

$$1 \longrightarrow \text{Inn } A \longrightarrow \text{LInn } A \xrightarrow{h_1} \text{Pic}(A,Z) \longrightarrow 0$$

is an exact sequence of groups and homomorphisms. □

Our results here can also be viewed as results about certain C^*-bundles. Generalizing from [13], we call ξ a *C_1^*-bundle over* X if ξ is an (H)C^*-bundle and $\Gamma_b(\xi)$ has an identity 1, so therefore $1(x)$ is the identity of $\xi(x)$ for each $x \in X$. If X is compact Hausdorff and ξ is a C_1^*-bundle over X such that

$$Z(\Gamma(\xi)) = C(X)1,$$

then the uniqueness in Theorem 2.6 insures $\xi_A \underset{X}{\cong} \xi$ where $A = \Gamma(\xi)$. We then write for simplicity $\text{LInn } \xi = \text{LInn } \Gamma(\xi)$, $\text{Inn } \xi = \text{Inn } \Gamma(\xi)$, etc. Thus by Theorem 3.10 and Proposition 3.11, $\text{KInn } \xi = \text{LInn } \xi$ and

$$1 \longrightarrow \text{Inn } \xi \longrightarrow \text{Aut}_X\xi \xrightarrow{h} H(\xi)$$

$$1 \longrightarrow \text{Inn } \xi \longrightarrow \text{LInn } \xi \xrightarrow{h_1} \text{Pic}(\xi, C(X)) \longrightarrow 0$$

are exact sequences. To get naturality of our exact sequences we must restrict attention to C_1^*-bundles for which each fibre has trivial center (henceforth denoted t.c.). We say ξ is a t.c. C_1^*-bundle over X if $Z(\xi(x)) = \mathbb{C}\cdot 1(x)$ for $x \in X$, and hence $C_b(X)\cdot 1 \subset Z(\Gamma_b(\xi))$ and $ev_x(Z(\Gamma_b(\xi))) = \mathbb{C}\cdot 1(x)$ for $x \in X$. Thus, if X is compact Hausdorff, then by Theorem 2.6 and Proposition 2.3, $C(X)\cdot 1 = Z(\Gamma(\xi))$, and if $A = \Gamma(\xi)$ with ξ a t.c. C_1^*-bundle over the compact Hausdorff space X, then by Theorem 2.6, $\xi_A \underset{X}{\cong} \xi$, so we can identify ξ_A with ξ. Suppose $f\colon Y \to X$ is a continuous map of compact Hausdorff spaces and ξ is a t.c. C_1^*-bundle over X. Then $f^*(\xi)$ is obviously a t.c. C_1^*-bundle over over Y and $f^*\colon\Gamma(\xi) \to \Gamma(f^*(\xi))$ is a unital $*$-homomorphism. Moreover, f^* also operates on $\mathrm{Inn}\,\xi$ and $\mathrm{Aut}_X\xi$ as well as $\mathrm{Pic}(\xi,C(X))$. Thus the pullback of $\alpha \in \mathrm{Aut}_X\xi$ is $f^*(\alpha) \in \mathrm{Aut}_Y f^*(\xi)$ and if $\alpha \in \mathrm{Inn}\,\xi$ then $f^*(\alpha) \in \mathrm{Inn}\,f^*(\xi)$ and in fact if $\alpha = \mathrm{ad}\,u$ for $u \in \Gamma(\xi)$ a unitary, then $f^*(u)$ is a unitary in $\Gamma(f^*(\xi))$ and

$$f^*(\alpha) = \mathrm{Ad}(f^*(u)).$$

If β is a natural Hilbert subbundle of ξ, then $f^*(\beta)$ is a natural Hilbert subbundle of $f^*(\xi)$ and if $\theta\colon \beta_1 \underset{X}{\cong} \beta_2$ then $f^*(\theta)\colon f^*(\beta_1) \underset{X}{\cong} f^*(\beta_2)$. Thus

$$f^*\colon \mathcal{H}(\xi) \longrightarrow \mathcal{H}(f^*(\xi))$$

is well-defined sending $[\beta] \in \mathcal{H}(\xi)$ to $[f^*(\beta)]$ which is in $\mathcal{H}(f^*(\xi))$. If $\alpha \in \mathrm{Aut}_X\xi$, then as already observed, if $x \in X$ and u is a unit vector in $\xi_\alpha(x)$, then u is unitary and $\alpha_x = \mathrm{Ad}(u)$. It follows that if $u \in \Gamma(\xi_\alpha|W)$ is a local unit vector field, then u is a local unitary section of ξ and $\alpha|W = \mathrm{Ad}(u)$; hence $f^*(\alpha)|W' = \mathrm{Ad}\, f_0^*(u)$, where $W' = f^{-1}(W)$ and $f_0\colon W' \to W$ is the map defined by f. Thus $f_0^*(u)$ is a section of $\xi_{f^*(\alpha)}|W'$, but $f_0^*(u) \in \Gamma(f^*(\xi_\alpha)|W')$ so that from this we see that

$$f^*(\xi_\alpha) \subset \xi_{f^*(\alpha)},$$

and hence $f^*(h(\alpha)) \leq h(f^*(\alpha))$, for $\alpha \in \mathrm{Aut}_X \xi$. But if $\alpha \in \mathrm{LInn}\, \xi$, then ξ_α is a line bundle so that $f^*(\xi_\alpha)$ is also, and hence as $\xi_{f^*(\alpha)}$ is a (0,1)-bundle, the previous inclusion is forced to be equality in this case. Thus if $\alpha \in \mathrm{LInn}\, \xi$, then $f^* h_1(\alpha) = h_1 f^*(\alpha)$ and hence we have a commutative diagram of short exact sequences

$$\begin{array}{ccccccccc} 1 & \longrightarrow & \mathrm{Inn}\, \xi & \longrightarrow & \mathrm{LInn}\, \xi & \xrightarrow{h_1} & \mathrm{Pic}(\xi, C(X)) & \longrightarrow & 0 \\ & & \downarrow f^* & & \downarrow f^* & & \downarrow f^* & & \\ 1 & \longrightarrow & \mathrm{Inn}\, f^*(\xi) & \longrightarrow & \mathrm{LInn}\, f^*(\xi) & \xrightarrow{h_1} & \mathrm{Pic}(f^*(\xi), C(Y)) & \longrightarrow & 0 \end{array},$$

showing that the short exact sequence behaves naturally with respect to f^*.

How to treat C*-algebras having no identity is not clear from our preceding developments. If A is a C*-algebra without identity and if $u \in M(A)$ is unitary then $\mathrm{Ad}(u)$ carries A into A so $\mathrm{Ad}(u)|A \in \mathrm{Aut}\, A$. Thus, it is reasonable to set

$$\text{Inn } A = \{(\text{Ad } u)|A : u \in U(M(A))\}$$

so that with $\text{Ad}_0 u = (\text{Ad } u)|A$,

$$\text{Ad}_0 : U(M(A)) \longrightarrow \text{Inn } A$$

is a group homomorphism. If $\text{Ad}(u) = \text{id}_A$, then $ua = au$ for all $a \in A$, and as $M(A) \subset A^{dd}$ and multiplication is separately ultra-weakly continuous, it follows that $u \in U(Z(M(A)))$, hence

$$1 \longrightarrow U(Z(M(A))) \longrightarrow U(M(A)) \xrightarrow{\text{Ad}_0} \text{Inn } A \longrightarrow 1$$

is a short exact sequence. If $\alpha \in \text{Aut } A$, then $\alpha^{dd} \in \text{Aut } A^{dd}$ and as

$$M(A) = \{b \in A^{dd} : bA \subset A \supset Ab\} ,$$

$M(A)$ is invariant under α^{dd} and defines $M(\alpha) \in \text{Aut } M(A)$ extending α. We define $Z = Z(M(A))$ and

$$\text{Aut}_Z A = \{\alpha \in \text{Aut } A : M(\alpha) \in \text{Aut}_Z M(A)\} .$$

Since $M(\text{Ad}_0 u) = \text{Ad } u$ for $u \in U(M(A))$, it follows that $\text{Inn } A$ is a subgroup of $\text{Aut}_Z A$ and is normal, as $\text{Inn } M(A)$ is normal in $\text{Aut } M(A)$, and $M: \text{Aut } A \to \text{Aut } M(A)$ is a group homomorphism, which is obviously injective.

We now define

$$\text{LInn } A = M^{-1}(\text{LInn } M(A)) \; ;$$

$$\text{PInn } A = M^{-1}(\text{PInn } M(A)) \; ,$$

$$\mathcal{H}(A) = \mathcal{H}(M(A)) \; .$$

$$\text{Pic}(A,Z) = h'(\text{LInn } A) \subset \text{Pic}(M(A),Z)$$

where $h' = h \circ M\colon \text{Aut}_Z A \to \mathcal{H}(M(A))$. Thus $\text{LInn } A \subset \text{PInn } A$ and $\text{PInn } A \subset \text{Aut}_Z A$. We then have a commutative diagram

$$\begin{array}{ccccccc}
1 & \longrightarrow & \text{Inn } A & \longrightarrow & \text{Aut}_Z A & \xrightarrow{h'} & \mathcal{H}(A) \\
 & & \downarrow M & & \downarrow M & & \| \\
1 & \longrightarrow & \text{Inn } M(A) & \longrightarrow & \text{Aut}_Z M(A) & \xrightarrow{h} & \mathcal{H}(M(A))
\end{array}$$

whose bottom row is exact by Theorem 3.6, and whose top row is easily now seen exact from the above definitions and the fact that $M\colon \text{Aut } A \to \text{Aut } M(A)$ is injective. But then by definition of $\text{Pic}(A,Z)$ we have the following commutative diagram with exact rows, where $h_1' = h'|\text{LInn } A$,

$$\begin{array}{ccccccccc}
1 & \longrightarrow & \text{Inn } A & \longrightarrow & \text{LInn } A & \xrightarrow{h_1'} & \text{Pic}(A,Z) & \longrightarrow & 0 \\
 & & \downarrow M & & \downarrow M & & \downarrow & & \\
1 & \longrightarrow & \text{Inn } M(A) & \longrightarrow & \text{LInn } M(A) & \xrightarrow{h_1} & \text{Pic}(M(A),Z) & \longrightarrow & 0
\end{array}$$

Since $\mathrm{Pic}(M(A),Z) \subset \mathrm{Pic}\, Z$, it follows that $\mathrm{Pic}(A,Z) \subset \mathrm{Pic}\, Z$, where $\mathrm{Pic}\, Z$ is the classical Picard group as described at the end of Chapter 2. Moreover, by our definitions, in the inclusion

$$\mathrm{LInn}\, A \subset \mathrm{PInn}\, A ,$$

an equality can be concluded from the corresponding equality for $M(A)$. Unfortunately, it does not seem clear that there is any reasonable way to determine which elements of $\mathrm{Pic}(M(A),Z)$ belong to $\mathrm{Pic}(A,Z)$. However, when $A = M(A)$, we will see in the next chapter that there is a useful way to characterize the members of $\mathrm{Pic}\, Z$ which belong to $\mathrm{Pic}(A,Z)$.

4 Relation to the classical Picard group

Let X be a compact Hausdorff space and set $Z = C(X)$. As in the preceding chapter, all Hilbert bundles considered are (F)Hilbert bundles. We now recall from the end of Chapter 2 that $\mathrm{Pic}(X) \cong \mathrm{Pic}\ Z \cong \check{H}^2(X; \mathbb{Z})$ is the classical Picard group. As in the preceding chapter, let A be a C*-algebra with identity and center Z. Then since $\mathrm{Pic}(A,Z) \subset \mathrm{Pic}(X)$ we may regard

$$\mathrm{Pic}(A,Z) \subset \mathrm{Pic}\ Z,$$

as the isomorphism $\mathrm{Pic}(X) \cong \mathrm{Pic}\ Z$ is also given by the section functor via Theorem 2.6. It is easy to see that $\mathrm{Pic}(Z,Z)$ is the trivial group so that $\mathrm{Pic}(A,Z)$ is somehow wrapped up in the non-commutativity of A. In this chapter we shall concentrate mainly on $\mathrm{Pic}(A,Z)$, having established its use in the preceding chapter.

First recall that if Y is any space and if F is contractible, then for $C \subset Y$, any $f: C \to F$ which has an extension to a halo of C in fact has an extension to all of Y, as shown for instance in [9, Fact 1, page 311]. Thus the trivial bundle $\varepsilon(Y;F)$ has the SEP as in [6], so by [7, Theorem 2.7], any locally trivial bundle with fibre F over a paracompact space must also have the SEP, and in particular, must admit a global section. These facts will be used without explicit mention in what follows.

We shall proceed to answer the question as to which $[\ell] \in \mathrm{Pic}(X)$ belong to $\mathrm{Pic}(A,Z)$. If K is a Hilbert space and B is a C*-algebra with identity 1, call $\phi: K \to B$ as H*-<u>linear</u> <u>map</u> if ϕ is linear and

(1) $\phi(K)\,\phi(K)^* \subset Z(B)$,

(2) for $a, b \in K$, $\phi(b)^*\phi(a) = \langle a|b\rangle 1$. (4.1)

From (4.1) it follows that ϕ is isometric, $\dim K \leq 1$, and $\phi(u)$ is a unitary of B whenever $u \in K$ is a unit vector. Indeed, ϕ is obviously isometric by (2) of (4.1) and if $v = \phi(u)$, then $v^*v = 1$. But $vv^* \in Z$ by (1) of (4.1) so that $1 = v^*(vv^*)v = (vv^*)v^*v = vv^*$, showing that $v = \phi(u)$ is unitary. If $s \in K$, then

$$\phi(s) = vv^*\phi(s) = v\langle s|u\rangle = \phi(\langle s|u\rangle u)$$

so as ϕ is isometric we have $\dim K = 1$ if $K \neq 0$. Let $H^*(K,B)$ denote the set of H^*-linear maps of K into B with the operator norm topology. If $u \in K$ is a unit vector, then it is clear that evaluation at u is an isometric isomorphism of $L(K,B)$ onto B carrying $H^*(K,B)$ onto $U(B)$, the unitary group of B with the norm topology. Suppose now that $[\ell] \in \mathrm{Pic}(X)$. Then using [11, Section 1] there is a subbundle $H^*(\ell,\xi_A)$ of $L(\ell,\xi_A)$ such that $[H^*(\ell,\xi_A)](x) = H^*(\ell(x),\xi_A(x))$ for each $x \in X$. Moreover, under the natural isomorphism $\Gamma(L(\ell, \xi_A)) = \mathrm{Hom}_X(\ell, \xi_A)$, sections of $H^*(\ell,\xi_A)$ correspond to Banach bundle maps which are H^*-linear on fibres. As $H^*(K,B) \subset U(K,B)$, it follows $H^*(\ell, \xi_A) \subset U(\ell, \xi_A)$ and hence if $\phi \in \Gamma(H^*(\ell,\xi_A))$, then $\phi(\ell)$ is a natural Hilbert subbundle of ξ_A and $\ell \cong \phi(\ell)$ over X. But this means $[\ell] \in \mathrm{Pic}(A,Z)$ if and only if $H^*(\ell,\xi_A)$ has a section. Also, as noted in [13, Section 1] the formation of $L(\ell, \xi_A)$ is natural with respect to pullbacks and restrictions to subsets of X, so the same is true of $H^*(\ell, \xi_A)$. As the referee kindly informed us, the image $\phi(K)$ of an H^*-linear map $\phi : K \to B$ is an example of a "subspace of isometries" according to [9]. In fact our natural Hilbert Z-submodules form a natural generalization of their notion from the setting of Hilbert spaces to the level of Hilbert modules.

Now, if ξ is locally trivial with fibre B, then ξ_A is a fibre bundle with fibre B and (see Chapter 1) group $\mathrm{Aut}^s B$, the group of *-automorphisms of B under the strong operator topology. It follows that in this case, $H^*(\ell, \xi_A)$ is locally trivial with fibre $H^*(\mathbb{C},B) = U(B)$ and group $S^1 \times \mathrm{Aut}^s B$. In the following theorem we summarize the main points of the preceding discussion.

THEOREM 4.1. *For* $[\ell] \in \mathrm{Pic}(X)$ *there is a bundle* $H^*(\ell, \xi_A)$ *over* X *with the following properties:*

(1) *the fibre of* $H^*(\ell, \xi_A)$ *over the point* $x \in X$ *is homeomorphic to* $U(\xi_A(x))$;

(2) *If* ξ_A *is locally trivial over* $C \subset X$ *with fibre* B, *then* $H^*(\ell, \xi_A)|C$ *is a fibre bundle with fibre* U(B) *and group* $S^1 \times \mathrm{Aut}^{(s)} B$;

(3) $[\ell] \in \mathrm{Pic}(A,Z)$ *if and only if* $H^*(\ell, \xi_A)$ *has a section.* □

COROLLARY 4.2. *If* ξ_A *is locally trivial with fibre* B *and if* U(B) *is contractible, then*

$$\mathrm{Pic}(A,Z) = \mathrm{Pic}\, Z. \qquad \square$$

The results of Phillips and Raeburn can now be quickly deduced. Suppose that $A = \Gamma(\eta)$ where η is a fibre bundle over X with fibre B a C*-algebra with identity 1 and with $Z(B) = \mathbb{C}\cdot 1$ and U(B) contractible. Then $Z(A) = C(X)$ and $\xi_A \underset{X}{\cong} \eta$ by Theorem 2.6, so Corollary 4.2 can be applied together with Theorem 3.10 and proposition 3.11 to give an exact sequence

$$1 \to \mathrm{Inn}\, A \to \mathrm{LInn}\, A \to \check{H}^2(X;\mathbb{Z}) \to 0.$$

The additional hypotheses on B used by Phillips and Raeburn are necessary to obtain $\Pi\mathrm{Inn}A = \mathrm{LInn}\ A$, where $\Pi\mathrm{Inn}\ A = \{\alpha \in \mathrm{Aut}\ A : \alpha^{dd} \in \mathrm{Inn}\ A^{dd}\}$, but once that is obtained, our analysis shows that the exact sequence is more general. Moreover, the analysis required to show $\Pi\mathrm{Inn}\ A$ coincident with $\mathrm{LInn}\ A$ proceeds purely locally so is a question only about the trivial bundle with fibre B, as shown in [36].

The advantage of a result like Theorem 4.1 is that it can be used to deal with algebras A where ξ_A is for instance a C*-bundle of the type considered in [13, MAIN THEOREM] over some specified base X whose topology is reasonably known. The strategy is to break X up into parts over which ξ_A is locally trivial and then successively apply the standard obstruction theory.

In particular, we have the next result as an application of the latter procedure.

COROLLARY 4.3. *Suppose that* $X = SY$ *is a suspension of a compact space* Y *with upper cone* C^+ *and lower cone* C^-, *so* $C^+ \cup C^- = X$, $C^+ \cap C^- = Y$ *and suppose that* C^- *and* $X\backslash C^-$ *are both contractible. If* ξ_A *is an* (F)C*-*bundle over* X *which is locally trivial over* C^- *with fibre* F *and locally trivial over* $X\backslash C^-$ *with fibre* B, *and if* $[Y,U(B)] = 0$, *then*

$$\mathrm{Pic}(A,Z) = \mathrm{Pic}\ Z.$$

Proof. Let U be a compact cone in X containing C^- in its interior with $U\backslash C^- = Y \times J$ for some interval J. Using results of [12] we can initially embed $\varepsilon(U;F)$ in $\xi_A|U$ and as ℓ is trivial over U, $H^*(\ell|U, \varepsilon(U;F))$ has a section. As the embedding $\varepsilon(U;F) < \xi_A|U$ is unital, it follows that $H^*(\ell|U, \varepsilon(U;F))$ is a subbundle of $H^*(\ell, \xi_A)|U$, hence $H^*(\ell,\xi_A)$ has a section over U, say s. Now $H^*(\ell, \xi_A)|X\backslash C^-$ is trivial with fibre $U(B)$ and the principal part of $s|U\backslash C^-$ corresponds to a map $U\backslash C^- \to U(B)$ which extends over all of X as $U\backslash C^-$ is homotopic to Y and $[Y,U(B)] = 0$. But then s must also extend to a section of $H^*(\ell, \xi_A)$, hence $[\ell] \in \mathrm{Pic}(A,Z)$. □

Notice that for $X = SY$ we have $\mathrm{Pic}\ Z \cong \check{H}^2(X,\mathbb{Z}) \cong H^1(Y,\mathbb{Z})$ so that in the situation of corollary 4.3 we have $\mathrm{Pic}(A,Z) \cong H^1(Y,\mathbb{Z})$ which can obviously be non-trivial. In particular, even though $\mathrm{Pic}(A,Z)$ is contained in the torsion part of Pic Z if ξ_A is locally trivial with finite dimensional fibre, this need not be the case when finite dimensional fibres and infinite dimensional fibres occur together, a phenomenon similar to that exhibited by results of [13, Section 10].

A technique similar to that of corollary 4.3 gives the next result whose details we leave to the reader

COROLLARY 4.4 _Let_ ξ_A _be an_ (F)C*-_bundle over_ X, _suppose_ $C \subset X$ _is closed and contractible, suppose_ $\dim F < \infty$, _and suppose_ $\xi_A(x) = F$, _for each_ $x \in C$. _Suppose that_ $U(B)$ _is contractible and that_ $\xi_A|X\backslash C$ _is locally trivial with fibre_ B. _Then_

$$\mathrm{Pic}(A,Z) = \mathrm{Pic}\ Z. \qquad \square$$

Recently [16] it has been observed that if A is any stable C*-algebra and if $Z = Z(M(A))$, then $\mathrm{Pic}(A,Z) = \mathrm{Pic}\ Z$. We simply apply the following proposition.

PROPOSITION 4.5. *Let* A *and* B *be* C*-*algebras with* $M(A) \subset M(B)$ *and* $Z(M(A)) = Z(M(B)) = Z$. *Then* $\Delta(M(A))$ *is a subsemigroup of* $\Delta(M(B))$ *which contains the identity of* $\Delta(M(B))$, *and* $\mathrm{Pic}(A,Z)$ *is a subgroup of* $\mathrm{Pic}(B,Z)$.

Proof. Every Hilbert Z-submodule of $M(A)$ is already a Hilbert Z-submodule of $M(B)$. □

Thus, if A is stable, then $M(A) \cong M(A \otimes K)$ contains $M(A) \otimes M(K) = M(A) \otimes B(H)$ which contains $Z \otimes B(H)$ as a Z-subalgebra, so

$$\mathrm{Pic}\ Z = \mathrm{Pic}(Z \otimes B(H), Z) \subset \mathrm{Pic}(A,Z) \subset \mathrm{Pic}\ Z$$

where K is the C*-algebra of compact operators on the separable infinite dimensional Hilbert space H.

As first pointed out in [36], there has to be a relationship between the group Pic*A of [2] and our group $\mathrm{Pic}(A,Z)$. In fact the relationship has been explicitly worked out for the case where A has continuous trace [38] and the general case will be treated in [16].

References

1. H. Bass, "Algebraic K-Theory," W. A. Benjamin, Inc., New York, 1968.

2. L. G. Brown, P. Green, and M. A. Rieffel, Stable isomorphism and strong Morita equivalence of C*-algebras, Pacific J. of Math., Vol. 71, No. 2 (1977), 349-363.

3. J. Dauns and K. H. Hofmann, Representation of rings by sections, Mem. Amer. Math. Soc. 83(1968).

4. J. Dixmier, "Les C*-Algebres et leurs Representations," Gauthier-Villars, Paris, 1969.

5. J. Dixmier, Ideal center of a C*-algebra, Duke Math. J. 35 (1968), 375-382.

6. J. Dixmier and A. Douady, Champs continus d'espaces hilbertiennes et des C*-algebres, Bull. Soc. Math. France 91 (1963), 227-283.

7. A. Dold, Partitions of unity in the theory of fibrations, Ann. of Math. 78(1963), 233-255.

8. S. Doplicher and J. E. Roberts, Fields, Statistics and Non-Abelian Gauge Groups, Comm. Math. Phys. 28 (1972), 331-348.

9. M. J. Dupré, The classification of Hilbert bundles, Ph.D. Dissertation, University of Pennsylvania, August, 1972.

10. M. J. Dupré, Classifying Hilbert bundles, J. Functional Analysis 15 (1974), 244-278.

11. M. J. Dupré, Classifying Hilbert bundles. II, J. Functional Analysis 22 (1976), 295-322.

12. M. J. Dupré, Hilbert bundles with infinite dimensional fibres, in "Recent Advances in the Representation Theory of Rings and C*-algebras by Continuous Sections," pp. 165-176, Mem. Amer. Math. Soc. 148(1974).

13. M. J. Dupré, The classification and structure of C*-algebra bundles, Mem. Amer. Math. Soc. 222(1979).

14. M. J. Dupré, Hilbert modules as non-commutative Hilbert bundles. In Preparation.

15. M. J. Dupré and P. A. Fillmore, Triviality theorems for Hilbert modules, in "Proceedings of the 5th Roumanian Conference on Operatory Theory," Birkhauser Verlag, 1981, 71-79.

16. M. J. Dupré, J. Phillips and I. Raeburn, Picard groups for C*-algebras, in preparation.

17. G. A. Elliott and D. Olesen, A simple proof of the Dauns-Hofmann Theorem, Math. Scand. 34 (1974), 231-234.

18. J. M. G. Fell, The structure of algebras of operator fields, Acta. Math. 106(1961), 233-280.

19. J. M. G. Fell, An extension of Mackey's method to Banach *-algebraic bundles, Mem. Amer. Math. Soc. 90(1969).

20. J. M. G. Fell, "Induced Representations and Banach *-Algebraic Bundles," Lecture Notes in Mathematics No. 582, Springer-Verlag, Berlin/Heidelberg/New York, 1977.

21. G. Gierz, Bundles of Topological Vector Spaces and Their Duality, Lecture Notes in Math., No. 955, Springer-Verlag, 1982.

22. R. Godement, Theorie generale des sommes continus d'espaces de Banach, C. R. Acad. Sci. Paris 228(1949), 1321-1323.

23. R. Godement, Sur la theorie des representations unitaires, Ann. of Math. 53(1951), 68-124.

24. A. Grothendieck, "A General theory of Fibre Spaces with Structure Sheaf," Univ. of Kansas, Lawrence, Kansas, 1955.

25. K. H. Hofmann, Representation of algebras by continuous sections, Bull. Amer. Math. Soc. 78 (1972), 291-373.

26. K. H. Hofmann, "Banach Bundles," Darmstadt Notes, 1974.

27. K. H. Hofmann, Banach bundles and sheaves in the category of Banach spaces, preprint.

28. K. H. Hofmann and J. R. Liukkonen, Editors, Recent Advances in Representation Theory of Rings and C^*-Algebras by Continuous Sections, Mem. Amer. Math. Soc. 148 (1974).

29. D. Husemoller, "Fiber Bundles," McGraw-Hill, New York, 1966.

30. I. Kaplansky, The structure of certain operator algebras, Trans. Amer. Math. Soc. 70(1951), 219-255.

31. G. G. Kasparov, Hilbert C*-modules: theorems of Stinespring and Voiculescu, J. of Operator Theory, 4 (1980), 133-150.

32. F. Krauss, Structure Theory of C*-Algebras, Dissertation, Tulane University, 1973.

33. E. C. Lance, Automorphisms of certain operator algebras. Amer. J. Math. 91(1969), 160-174.

34. R.-Y. Lee, On the C*-algebras of operator fields, Indiana Univ. Math. J. 25(1976), 303-314.

35. W. L. Pachke, Inner product modules over B*-algebras, Trans. A.M.S., 182(1973), 443-468.

36. J. Phillips and I. Raeburn, Automorphisms of C*-algebras and second Cech cohomology, Indiana U. Math. J., Vol. 29 No. 6(1980), 799-822.

37. M. Pimsner, S. Popa, and D. Voiculescu, Homogeneous C*-extensions of C(X) O K(H), J. Operator Theory 1 (1979), 55-108.

38. I. Raeburn, On the Picard group of a continuous trace C*-algebra, preprint.

39. M. A. Rieffel, Induced representations of C*-algebras, Adv. in Math. 13(1974), 176-257.

40. M.-S.B. Smith, On automorphism groups of C*-algebras, Trans. Amer. Math. Soc. 152(1970), 623-648.

41. N. Steenrod, "The Topology of Fibre Bundles," Princeton Univ. Press, Princeton, N.J., 1951.

42. J. Varela, Duality of C*-algebras, in "Recent Advances in the Representation Theory of Rings and C*-algebras by Continuous Sections," pp. 97-108, Mem. Amer. Math. Soc. 148(1974).

List of symbols

Page

Page

Page

Page

Index